PULSE•2

report on a phenomenon

This publication and the exhibitions it accompanies were made possible by a grant from the David Bermant Foundation: Color, Light, Motion with additional generous support from the University Museum Council.

PULSE • 2

report on a phenomenon

Phyllis Plous

With essays by

Nancy Doll
Diane Shamash

University Art Museum Santa Barbara

Distributed by University of Washington Press
Seattle • London

Exhibition Locations

PULSE 2

7 July – 21 October 1990
University Art Museum
Santa Barbara

7 July – 5 October 1990
College of Creative Studies Gallery

7 July – 26 October
University Center Gallery

PULSE 2 / DOWNTOWN

1 August – 21 October
Channing Peake Gallery
Santa Barbara County Arts Commission

IMPROBABLE MACHINES

1 September – 28 October
Santa Barbara Museum of Art

RAUL RUIZ / THE EXPULSION OF THE MOORS

22 September – 27 October
Santa Barbara Contemporary Arts Forum

Library of Congress Cataloging-in-Publication Data

Pulse 2: report on a phenomenon / introduction by Phyllis Plous; essays by Nancy Doll and Diane Shamash.
76 pp. cm.
Catalog of an exhibition, University Art Museum, Santa Barbara;
ISBN 0-942006-19-4

1. Art: art and technology—United States—Exhibitions. 2. Pulse techniques (Electronics, Industrial materials) in art—Exhibitions. I. Plous, Phyllis. II. University of California, Santa Barbara, University Art Museum. III. Title: PULSE two.
N72. T4P85 1990
709'.73'07479491—dc20
Library of Congress Catalogue Card Number 9-36493 CIP

Performance Locations

10 – 26 October 1990
College of Creative Studies Gallery

Richard Dunlap, *History of Animals*
10, 11 October

David Mahler, *Listener in Residence & The Art of Tape*
16, 17, 18 October

Mimi Garrard / James Seawright, *Walking on Gravel*
22 – 26 October

Photograph Credits

Jeff Brouws
Bill Buchen, New York
Geoffrey Clements, New York
D. James Dee, New York
William Dewey, Santa Barbara
T. Charles Erickson, New Haven
Ralph Gabriner, New York
Hansen/Mayer, Boston
Jürgen Hilmer, Santa Barbara
Bill Jacobson Studio, New York
James P. McCoy, New York
James Seawright, New York
Squidds & Nunns, Los Angeles
Visions West, Victoria, B.C.

Designed by Bollinger & Peters, Santa Barbara
Printed in an edition of 3,000 by Typecraft, Inc., Pasadena

Table of Contents

Lenders

Art Gallery of Greater Victoria, Vancouver
Art for Public Space, Inc., Harrison, New York
Steve Barry
David and Carol Becker
Barry and Gail Berkus
David W. Bermant
Bermant Enterprises, Ltd., Harrison, New York
Eli Broad Family Foundation, Santa Monica
Bill and Mary Buchen
Martin and Sharleen Cohen
Paula Cooper Gallery, New York
Steven Cortright
Jack Dollhausen
Richard Dunlap
Bill Fontana
Mimi Garrard
Dorothy Goldeen Gallery, Santa Monica
Michael Hayden
Brandyn Jones
Evan Lewis
Robert and Mary Looker
Luhring Augustine, New York
Luhring Augustine Hetzler, Santa Monica
David Mahler
Museum of Contemporary Art, Los Angeles
National Shopping Centers, Inc., Harrison, New York
Newport Harbor Art Museum, Newport Beach
Dennis Oppenheim
Liz Phillips
Private Collections
Max Protetch Gallery, New York
Carl and Judy Schlosberg
James Seawright
Moira and Alejandro Sina
Martin Sklar
Survival Research Laboratories, San Francisco
University Art Museum, Santa Barbara
Vancouver Art Gallery, British Columbia
Germans van Eck, New York
Ted Victoria
John Weber Gallery, New York
Whitney Museum of American Art, New York

Foreword

Since its founding thirty years ago, the University Art Museum has been committed to the exhibition and interpretation of contemporary art. With PULSE 2, an acronym for people using light, sound, energy, curator Phyllis Plous continues her explorations of advanced, often not widely accepted art forms. Most of these exhibitions have been mounted in the Museum's galleries in which, necessarily, limited numbers of works and few truly massive ones have been shown. In PULSE 2, however, we have collaborated with two other galleries on campus, three additional organizations in downtown Santa Barbara, and used several outdoor sites, for what is easily the largest and most ambitious exhibition ever undertaken by this Museum. It is a credit to Mrs. Plous' organizational ability and the gracious cooperation of the entire Santa Barbara art community that PULSE 2 has come into being.

The broad geographical and conceptual ground covered by PULSE 2 embraces diverse themes, processes, and media. There are pieces featuring sound with mysterious sources, while others can be heard only if personally activated. The use of natural or man-generated energy in nearly all the art presented here also implies the importance of time in experiencing them. Spatial, audio, and light level or color changes in this art call for a cataloguing technique other than conventional printed analyses and still pictures. We have therefore produced a video catalogue as a vital complementary adjunct to this one. Together these permanent records fail, however, to record to our full satisfaction the dynamic, evanescent, and subtle qualities of the art assembled under the heading of PULSE 2.

In addition to thanking the museum staff for service well beyond the call of duty, I would like to join Phyllis Plous in thanking the lenders to and supporters of the exhibition. Foremost among them by far is David Bermant, in several ways the inspirational force behind PULSE 2. Without the generous support of the David Bermant Foundation: Color, Light, Motion, the exhibition could not have been mounted.

Herbert M. Cole
Acting Director

Acknowledgments

PULSE 2 marks a milestone both for the University Art Museum and the field of contemporary art as a whole. It constitutes a large-scale exhibition of light, sound and energy sculpture that combines traditional gallery installations with site-responsive public art projects. A celebratory event, it is also the first collaboration between several major art institutions in the Santa Barbara area. Clearly the artists, lenders, collaborators and financial supporters who have joined with us in the PULSE 2 project had the vision to perceive the exhibition as a serious investigation of artistic innovations in sculpture. As such, we hope the exhibition and its catalogue will contribute to a fuller understanding of the individual works in the exhibition and will engage the interest of a larger audience in these and other artworks that deal with developing technologies. Too often since the late 1960s, interactive sculpture has been lumped together under the rubric of "kinetic art" and dismissed as amusing, *déja vu* gadgetry.

Early in its organizational phase, David Bermant, a highly motivated, passionate collector of interactive sculpture, became an active supporter of the Museum's endeavor. In fact, the exhibition and its catalogue received the commitment, support and enthusiasm of many individuals and several institutions. Our foremost appreciation goes to the artists for making objects and installations of compelling interest. Their participatory assistance as we organized the exhibition and their generosity in sharing their time were enormous.

Second only to the artists in making this project possible is the diverse group of lenders. Particular gratitude goes to Mr. Bermant, founder, and James Seawright, president, of the David Bermant Foundation: Color, Light, Motion for their encouragement of this undertaking. To the private collectors, public institutions, foundations and galleries from whom we have borrowed works, we are deeply appreciative for their willingness to part temporarily with their sculptures. Their names are listed at the beginning of this catalogue. For their interest, support and information, I am also grateful to Tom Finkelpearl, Ellen O'Donnell, Karen Amiel, Richard Armstrong, Pamela Auchincloss, Bill and Mary Buchen, Richard Dunlap, Susan Hopmans, Milton Komisar, Stuart Regen, Diane Shamash, Jan Strayer and Malinda Wyatt.

We are delighted that the University Art Museum is sharing the exhibition in collaboration with several other campus and Santa Barbara institutions. For their interest in our project since its beginning and their support throughout the past three years, I want to thank Dan Connally, director, College of Creative Studies Gallery; Alan Kirby, director, University Center; Leigh Oppenheimer and Brenda Arechiga, the successive Associated Students' directors of the University Center Gallery; Nancy Doll, curator of twentieth century art at the Santa Barbara Museum of Art and organizer of the IMPROBABLE MACHINES exhibition; Maria de Herrerra, director of visual art for the Santa Barbara County Arts Commission and the Visual Art in Public Places Committee of the City of Santa Barbara and organizer of the PULSE 2 / DOWNTOWN exhibition and the downtown site installations; and Betty Klausner, director, Santa Barbara Contemporary Arts Forum.

On the campus itself, we have had consistent support and encouragement from Chancellor Barbara Uehling who has installed François and Bernard Baschet's Fountain in the garden entry of her campus residence; the Chancellor's Physical Planning Committee; Professors Leslie Wilson, Adrian Wenner and Stephen Rothstein, Biological Sciences; Larry Parsons and the Department of Health and Safety; Jay Sullivan who coordinated all activities with Facilities Management's helpful staff; Tye Simpson, campus planner for the Chancellor's office; Juan Betranena, planner for Facilities Management; and the Communications Services Department.

I would also like to thank Diane Shamash, Program Coordinator for the Seattle Arts Commission and curator Nancy Doll of the Santa Barbara Museum of Art who have enriched the catalogue with thought-provoking essays. The catalogue has benefited from Jürgen Hilmer's photographs of a number of the installations and objects as they appear in the exhibition. They have become, of course, the very important heart of this record. Laina Long's thirty-minute videotape, incorporating artists' site interviews, along with video coverage of the objects in the gallery spaces, is a carefully honed catalogue companion. It has been a pleasure to work with Gary Bollinger and Ginny Brush of Bollinger and Peters Design, who designed and coordinated the production of this catalogue with skill, good humor and infinite patience. As has so often been the case, John Daniel has provided detailed, sensitive editing.

The credit for seeing the exhibition and catalogue projects to completion belongs to the entire staff of this Museum. From the moment he assumed his post, acting director Herbert M. Cole endorsed the exhibition and shouldered some of its administrative responsibilities. Successful problem-solving of the exhibition's myriad challenges, presented by a wild variety of intricate objects and site-responsive installations, was carried out by exhibition designer, Paul Prince, assisted by preparator Rollin Fortier and their tireless crew as well as guest installers Chris Hopmans and Bruce Gluck. Together they provided ideal situations in three university gallery spaces for the objects and a series of installations, as well as a number of public art projects, each complex individually and in combination even more so.

My thanks to curatorial assistant Brian Parshall, who provided intelligent, assiduous support for several aspects of the project from coordination of many organizational details to research and preparation for sections of the catalogue. Curatorial assistants Rick Orr and Jackson Dodge furnished indispensable help in the early research and preparation for the exhibition. The newest member of our staff, Gary Todd, prepared the manuscript and, with keen attention to detail, became an invaluable aide in proofing it. Registrar Sandra Rushing and Andrea Pearson, assistant registrar, both recent members of our staff, pulled together transportation and insurance arrangements for this complex exhibition in exemplary fashion and record time. Sharon Major's diligent efforts in drawing the public's attention to the existence of PULSE 2 in Santa Barbara is greatly appreciated as are the special education and outreach activities conducted by Corinne Horowitz. I would also like to thank Roberta Cook, Barbara Vilander, staff members in the Museum's office, and Carla Lease, head guard.

This project has been an expensive and difficult undertaking. In funding PULSE 2, the David Bermant Foundation: Color, Light, Motion and the University Museum Council have demonstrated their belief in sculpture that uses developing technologies as a significant area of contemporary art, both within conventional gallery spaces as well as in the public domain. We are grateful to them.—*PBP*

Marcel Duchamp
Rotoreliefs
1935-1953

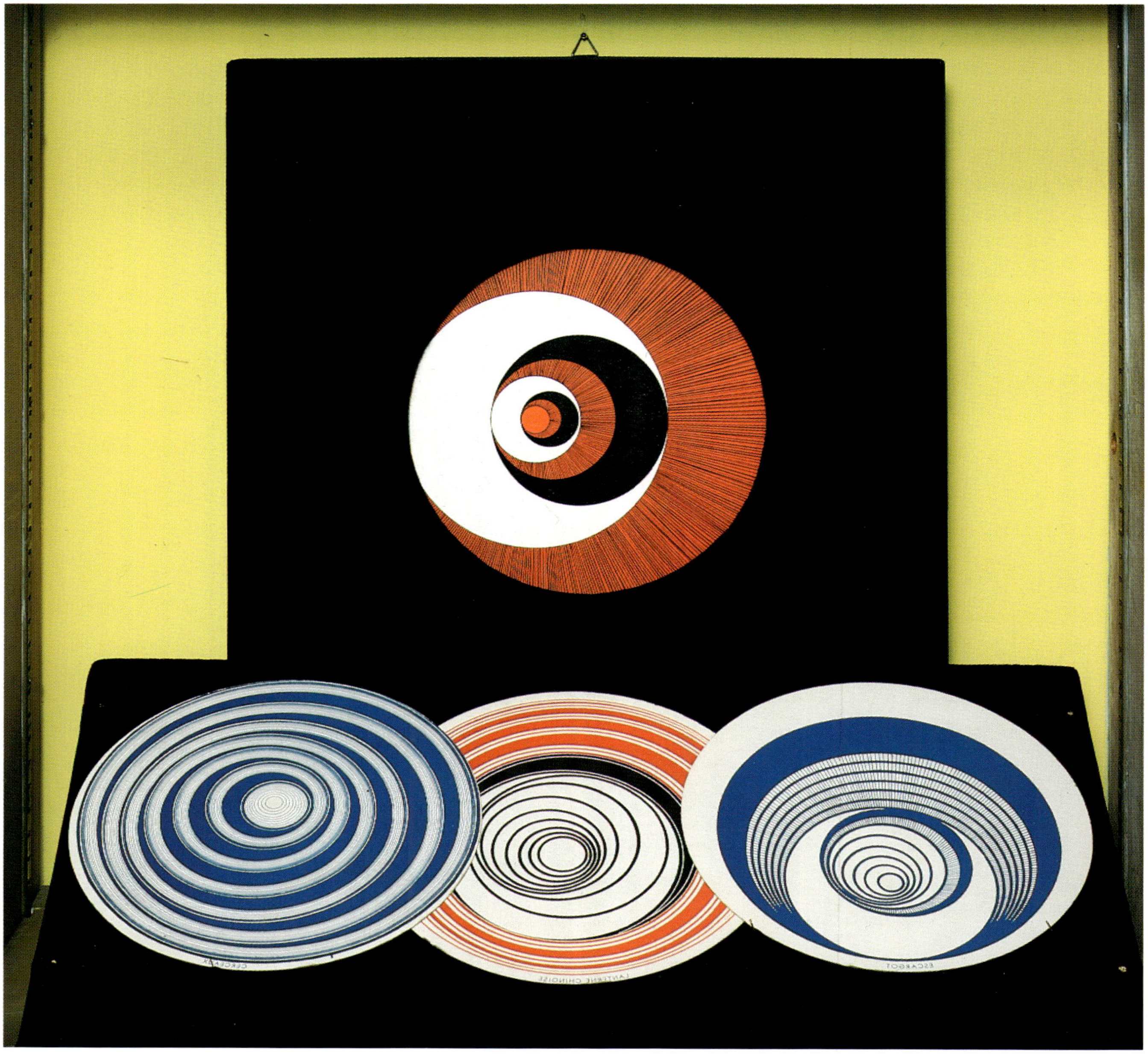

Report on a Phenomenon

Phyllis Plous

> *By the turn of this century, nothing, absolutely nothing, was on certain ground; everything was being challenged; this is the main fact of our lives.*
> —Alfred North Whitehead

PULSE 2 offers an overview of the relationship between technology and sculpture, one in which politics and metaphysics are equally present. To give as full an account as possible of sculptors who work with developing technologies to express their pre-eminent concerns, several interrelated goals have been established. The first is to present a carefully selected group of objects and installations ranging from some of the best vanguard work made in this sphere during recent years to works that contrast by dint of their delightful frivolity, for which the field is also well known. It seemed equally important to include a brief review of earlier twentieth-century precedents and attitudes rather than organizing the exhibition as a perpetually new mix of undifferentiated works. In this way the show charts a complex, often brash and undeniably intense history of intriguing efforts to develop new avenues for sculptural discussion. Priority was given to the placement of visually and technically sophisticated works alongside evocative, low-tech sculptures and to the presentation of several contemporary performances. Simultaneously, site-responsive public works were located in Santa Barbara's university campus and community. The exhibition is an attempt to discuss what it is that artists do with the available technology at hand today and why they do it. Art is what artists do, but we need to remember that artists don't always do what they say.

Western society moved with overwhelming speed from an era which substituted machines for muscle to a post-industrial period dominated by electronic and chemical devices that can, for instance, simulate the processes of the brain and the nervous system. By concentrating attention on artists who make widely varied uses of current technologies and their less sophisticated antecedents, specific artistic strategies and an inherent matrix of meanings can be readily compared, helping to clarify not only the similarities and differences between the works, but also their place in contemporary art.

As the Sphinx and the Great Pyramids of Giza issued from the cultural triad of art, science and technology so, in the Renaissance, new findings in the science of perspective had seminal influences on art. For the last eight decades, overwhelming advancements in the areas of science and mathematics have rapidly altered our artistic and cultural framework. There are curious similarities between Leonardo da Vinci, who, in a systematic manner, attacked so many branches of knowledge, and Marcel Duchamp who in attacking the notion of originality, played a large role in changing the condition of art in the twentieth century. Where Leonardo anticipated many later discoveries in anatomy, aeronautics and several other fields in addition to being one of Italy's greatest artists, Duchamp, an indispensable skeptic, functioned as a radical innovator for art. He introduced the concept that in the future viewers would need to abandon their passive role and become actual participants, thereby completing the work of art.

Duchamp's ubiquitous influence provided a fulcrum by which countless artists have redefined themselves, reversing the normal succession of events that usually forms around a significant body of work. In general, an artist has most influence on the art of his contemporaries and immediate successors, the power of his own work to stimulate younger artists becoming weaker as it reaches further into history. Picasso is a good example in that he affected entire generations of artists in the decades from 1910-1950 but has not figured prominently as an influence since then. Consider that there have been more art movements in the last seven or eight decades of this century than in the seven centuries preceding it. The logical consequences of the enigmatic Duchamp's speculative attitudes have influenced the attitudes of an impressive number of those movements, including much of the work in this exhibition.

Still, Duchamp's experiments in motion, concluding with the *Rotoreliefs*, 1935-1953 (no. 23), remain ambiguous. The mechanical aspects of his depersonalized devices were secondary to their metaphysical and aesthetic implications and were used to support his experiments with space, time and optical illusion. After the first shock that his earliest works produced, Duchamp moved on to investigate other aspects of the fourth dimension, notably chance and change, presenting strategic tensions between his psychologically charged, erotic metaphors and his disembodied conceptual questions. In retrospect, he emerges as a prophet of the concerns of twentieth-century artists from 1913 onward and especially of recent artists. Just as he periodically dismantled "art" in the classical meaning of the term by means of subversion and paradox, Duchamp smiled at art history while making certain his art ended up in museums.

The intellectual and psychic origins of the works in this exhibition lie in the unparalleled amount of scientific knowledge, cultural discoveries and rejuvenation that developed in the twentieth century as geometry, science, art and technology underwent linked transformations. Before 1905, when Albert Einstein produced his earliest theory of relativity, artists had been guided by their intuition and by reliance on both Euclidian geometry and that predictable, clock-like universe described by Newton.

Since the late teens and early twenties, artists have increasingly enmeshed themselves in non-static and nonlinear discoveries moving from such concepts as relativity to indeterminacy, quantum and chaos theories. These dominant forces have precipitated rapid changes in global culture, and on the brink of the twenty-first century artists are meditating on randomness, fractals and the order of chaos. They make searching inquiries into electronics and the new chemistry which in turn directs their concerns toward the physical, biological and cultural realms of contemporary life.

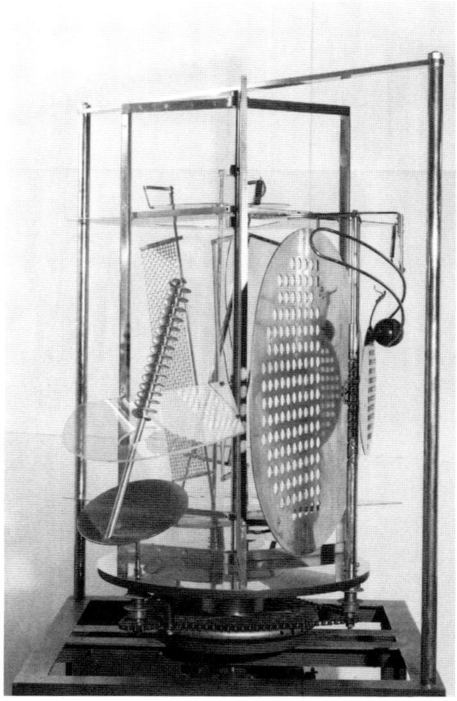

László Maholy-Nagy
Light-Space Modulator
1923

Pivotal to developing the possibilities for a future kinetic art that would incorporate real time or experienced time were Duchamp, Naum Gabo, László Moholy-Nagy and Alexander Calder. It was in 1920, that Duchamp made the first mecanomorphic sculpture, *Rotary Glass Plates*, and Gabo produced his geometric *Kinetic Construction*. In 1922, Moholy-Nagy followed with the formulation of a theory of kinetic art, defined as the "relationships of energies" rather than of forms. Both his theory and his sculpture, the startling *Light-Space Modulator*, 1922-1930 (no. 53), predicted theatricality in the visual arts and foresaw the day when art would be an event unfolding before the viewer as "vision in motion." In the 1930s, Calder, too, began to produce "virtual volume" with a simple vocabulary of balanced and counter-balanced elements that moved in a slowly spinning temporal sequence with an acrobatic playfulness. But these four progenitors realized that current technology was too cumbersome to carry out their concepts of motion in art and that the evolution of temporal art was going to require greater technological advances. They withdrew.

Then, for almost three decades, with the exception of Calder's hand- and wind-driven mobiles, kinetics went underground. When it re-emerged during the mid-1950s, there was a deep ambivalence concerning technology. On the one hand there was the pessimism carried over from the depression of the 1930s and the dropping of the bombs on Hiroshima and Nagasaki, and on the other hand, in America more than in Europe, there was an affluent, rather naive optimism regarding human values and man as the creator of machines. There was less optimism about technology itself.

Nonetheless, the new technological advances made during the period of World War II had become available to isolated artists (particularly in America) and various collective groups (especially in Europe) who responded to the dynamics of change and took up the practice of kinetic art. The widely ranging interests of these artists and groups and the diversity of media used in their art made a definition of their concerns difficult. A European-wide artistic ideology known as La Nouvelle Tendence began to spread across the continent as early as 1956, and in 1961 La Nouvelle Tendence was formally organized in Zagreb as an international movement concerned with the primacy of research, depersonalization of art, open communication and collective work. It opted for a flexible kineticism of mechanical relationships but left open the possibility for immateriality, luminosity and nature. Presenting a contradiction in terms, it promoted the idea that a rationally behaving machine could function, both in actuality and artistically, if injected with elements of imprecision and irrationality.

A lot of related activity was also generated in the United States in the late 1960s with the formation of Experiments in Art and Technology, Inc. E.A.T. was a loosely organized, interdisciplinary group that attempted to act as catalyst and match-

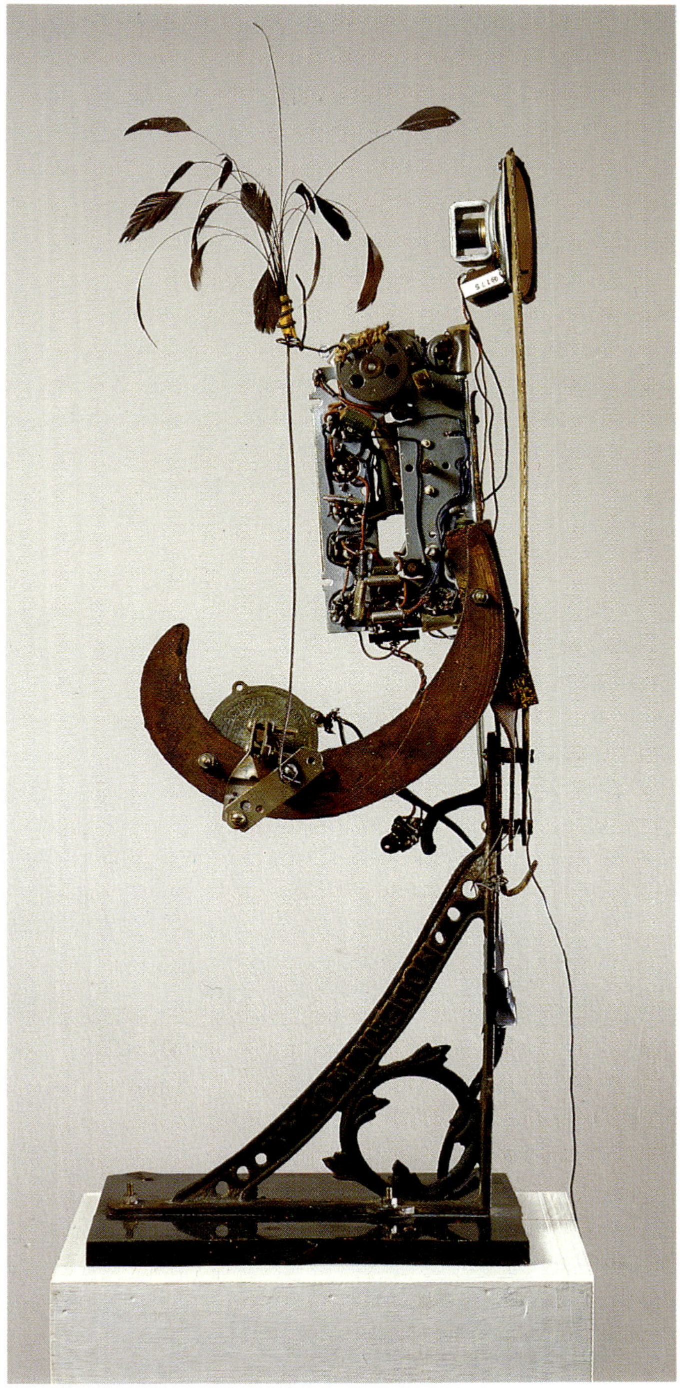

Jean Tinguely
Tokyo Gal
1967

maker, creating professional and social rapport between artists and engineers, the art world and industry. From 1967 to 1971, Maurice Tuchman carried out a related program of collaboration between artists and industrial firms which culminated in the *Art and Technology* exhibition at the Los Angeles County Museum of Art in 1971. R. Buckminster Fuller provided innovative architectural projects and insights concerning synergy vs. entropy and also pointed out the need to cease squandering our resources and consider the impact of our racing technological production in relation to the ecology of the planet. During the same period, Gyorgy Kepes introduced another interdisciplinary program when he brought numbers of artists from all media to M.I.T.'s Center for Advanced Visual Studies to participate in urban design, electronic art and sculpture.

Although the kinetic movement drew considerable public response in Europe, American response was reasonably cool. Museums, critics and collectors frowned on a problematic movement that involved overwhelming technical and financial difficulties. Defending kinetic art in Udo Kultermann's 1960 catalogue, *Monochrome Malerei*, Enrico Castellani stated that "An artistic tendency which, due to its very essence, is impossible to define in terms of pre-existing schools or movements is certainly not a phenomenon which can be ignored. . ."[1] At last in 1965, George Rickey, one of the movement's leading sculptors and its most articulate spokesperson, listed kinetic art's fundamental concerns. Bypassing traditional materials, he included: new ideas of the space-time continuum between two and three dimensions; use of micro-elements as creators of forms, tones and forces; non-Euclidian geometry; consideration of light and movement as expressive in themselves; the objectification of a work through mathematical relations of change, randomness and spectator participation; and optical phenomena where the artist uses the reaction to stimuli of the eye itself rather than what the eye surveys and the brain interprets.[2] Working with these ideas, the strongest of the kinetic artists were able to develop an art based on the acceptance of the invisible as well as the visible external forces that condition life. Generally they were works that both artist and viewer experienced as forms of performance.

Yet by the early 1970s, what had been regarded since the mid-1950s as provisional sculpture, by a large segment of the art world and the public, largely disappeared from public view again. Following 1968 a rapid wave of cultural change announced a new set of conditions for the post-industrial era. The Cold War continued and the two powers competed heavily for leadership in space exploration, man finally landed on the moon, and the American war in Vietnam alarmed the world. Those events became the source of social and environmental speculation and critical evaluation. With the word "war" always on our lips (the war on poverty, the war on drugs, etc.), society was also being transformed by the widespread use of computers and advances in communications technology. There was a cynical proliferation of war games in distant lands, and a growing menu of available drugs at home plus their increasing per capita use. These were events that could at least be recognized by art if not changed by it. But those who wanted art to solve the world's problems were in for confusion and disappointment. Art doesn't disturb the order of the contemporary world for the order of that world is its own order. As a result the artistic situation became extremely ambiguous in the period of the 1970s, a decade that acted out its time like a true if irrational evolution. In one sense artists came to understand a vision of process that involved a diverse array of "mutations," which altered and recombined the available pool of accessible forms and tolerated an expanded permissiveness long enough for notions with new information and criticality to take hold.

Then, beginning in the mid-1980s there was an acceleration of interest among artists who wanted to forge new alliances with developing systems and technologies. Much of the emphasis, however, was shifting toward a critique of these new systems while former, more positive notions of technological "progress" were giving way to suspicion. What do we mean when we speak about artists who make use of today's developing systems for sculptural works that cannot be defined as either genre, medium or movement? Given the plush amorality of the 1980s, has such work registered with the public as entertaining assemblages of expensive, moving hardware with a variable life expectancy? That's not likely. Rather artists and viewers are discussing interactive objects, installations and performances that spatially and optically preserve their own presence; or speculative and interrogatory works that reveal an account of the artist's making and his critical stance; or sculpture that may adopt an empirical approach but is as much concerned with perceptual psychology, concept, process and the participation of the spectator as it is with material veracities and structure.

Alan Rath, for example, makes interactive sculptures that embody the more insidious qualities of technology and their effect on human beings. He insinuates that technology reduces the importance of human activity to insignificance although he uses industrial and electronic media. Without being hackneyed, he raises metaphorical questions concerning our involvement in the technological race, concurrently redefining its high-tech parameters. He also deals with ambiguity in his play with tracks of time. As one critic observed, there is an emphasis on time: the real time of the work's technical performance, the lived time the viewer takes to observe it and the fictive time involved in each piece of programmed imagery.[3] In *Voyeur II*, 1989 (no. 63), there are ominous references to surveillance and paranoia which are chilling. *Introspective Sculpture*, 1988 (no. 61), is concerned with the kind of information, fueled by data retrieval, that has reached a dictatorial position vis à vis contemporary culture. Its obvious repetition of "1,0,1,0 . . ." on the cathode ray tube makes the point. Laced with humor, *Desk Top Breather*, 1989 (no. 62), is an electronic speaker that heaves like a human lung or a thumping heart under stress. Still, Rath's obvious references are not so blatant as to overwhelm the visual character of his works, which is always a risk for artists obsessed with technology.

In works such as those of Rebecca Horn, Liz Phillips or Michael Hardesty, the artists create machines or an environment of sensuous alertness where, in the Duchampian sense, viewers enact the subliminal activity or the metaphor that the work suggests. In being its audience, the viewer automatically becomes its performer. Phillips has a particularly contemporary understanding of our time, informed by an awareness of mutability. She is concerned with the natural environment in which we live and the conflict between universal time and individual time. The combined parts of *Soundtable I and II, (Wet and Dry Landscape)*, 1989-1990 (no. 59), are like specific notations serving as a continuum of energized information that bears upon itself just as one part bears upon another.

Rebecca Horn's art proceeds from a mixture of simplicity and complexity, a passive visual reading of her work suggest-

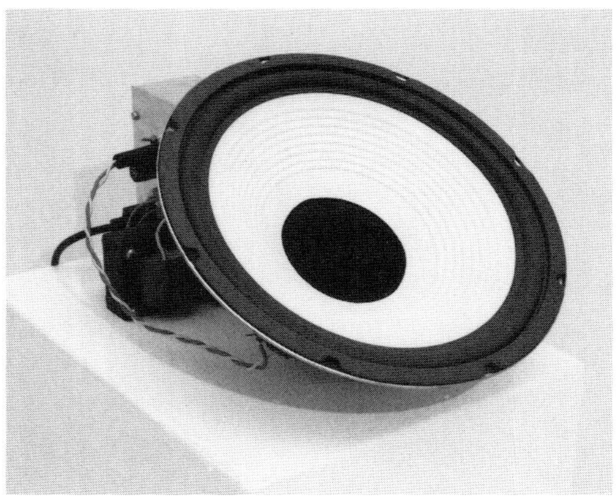

Alan Rath
Desktop Breather
1989

ing simplicity. But a complex one occurs to the viewer as sh/e attempts to situate it and identify with it. Horn's strength lies in her recognition of both the primacy of experience and the interdependence between a relationship with the self and one with the world. As with other motifs in her work, feathers signify an extension of the body's chemistry and fuse with the fleeting dynamic of touch itself. *Butterfly Machine*, 1986 (no. 34), features wings that open and close with seductive slowness at specific intervals. The artist's ability to draw our attention to that simple repeated movement through the use of anticipatory pauses creates anxieties within the viewer.

Not unlike the history of the California artist Charles Ray whose objects represent his performance-oriented, body sculptures from the 1970s and early 1980s, Horn's mechanized sculptural works represent her film/performances from the same period. Ray sometimes utilizes almost imperceptible motion or liquids as metaphors for bodily functions to provoke the viewer's experiential participation. Horn presents her universe, eroticized in a general way, obsessive, intensely personal and speculative, as testimony to her "desire to clarify, first to others and then to herself, the fascination of female fantasy."[4] Both artists reassert the physical fact of the body and the psychological fact of memory and intuition. Rather than seeking to critique the alienating conditions of mass-media consciousness, as did much art of the late 1980s, these artists attempt to gesture towards phenomena which remain irrefutably human.

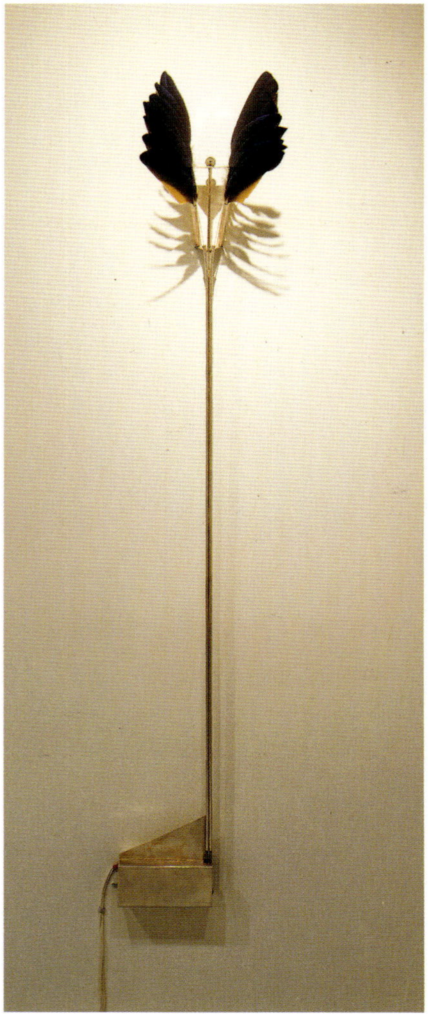

Rebecca Horn
Butterfly Machine
1986

Entirely different in its format and content is the more romantic and symbolic, low-tech work, *The Crew*, 1987 (no. 17), by Jim Clark. Clark pieced together discarded junk, wire and two lighted fluorescent tubes, previously dipped in paint, and placed them in a water-filled plastic tub supported by a simple wooden structure. More organic with its low-key materials than the high-tech electronic sculpture of Rath, Clark's work comments on the consumerist aesthetic of the 1980s.

Evan Lewis is interested in movement in nature, both in terms of the use of natural forces and in terms of movements which express his feelings about the surrounding world. His large *Quiet Storm*, 1990 (no. 44), is composed of two interrelated forms, each constructed with balanced elements which move with a heavy grace. Installed thirty feet above ground on the south facade of the campus dramatic arts building, it brings to mind wind-propelled farm equipment in a rural environment.

The work of Eric Orr is symbolic. In *Prime Matter VIII*, 1987 (no. 58), Orr is a pursuer of metaphysical visions who evokes alchemy and the earth's elements. The sculpture is one of a series of rectangular columns the artist began to make in the 1980s. When the work is fully operative, flame shoots up the center of the column as a miniature cascade of water runs and slides downs its sides.

In order to keep his camera obscura techniques subject to realms of fantasy and desire, Ted Victoria maintains a simple technological level of operation. In *Hey Pete, Heads or Tails?*, 1989 (no. 84), Victoria has created a surreal set-up. Two slowly spinning baseballs, placed within a flat cabinet, are projected by means of a system of light and mirrors onto translucent plastic so that they appear to float. In a similar manner, a magic lantern takes an upside-down wine glass in *Fill it to the Brim*, 1990 (no. 85), and transforms it into a large translucent wall image of water shooting up and out of a glass which is right side up. *Fill it to the Brim* and its companion piece, *Bottoms Up*, 1990 (no. 84), are installed respectively on exterior walls at the University Art Museum and the Santa Barbara Museum of Art.

With his robotic *House Plants*, 1984 (no. 71), James Seawright links the natural and electronic worlds. Referring to day- and night-blooming flowers, it's a behavioral sculpture that relies on computer programming designed to produce a patterned "personality." Here the visual and the behavioral are merged by the artist into one system, an integrated aesthetic unit responding to outside stimuli. The artist argues that the processes and principles of technology as a medium for art are as valid as wood, bronze or paint on canvas and that the conventions of understanding the medium are equally important. Seawright, who is a major spokesperson for artists using developing technologies, mentioned during a studio visit that the most remarkable thing to him about technological invention is that there is always a man behind it. His works reiterate art as a man-made construct. Seawright also believes that technology in the hands of an artist has the capacity to speak radically, and that it's possible to integrate the "material" of technology with the expression of a new content rather than expressing old ideas or traditional associations.

The foregoing artists deal with issues, supported by ever-advancing technologies, that are evidence of a striking change of direction since the early 1970s. However, the emphasis had been different in the 1960s when, with Europe in the lead, ideas about light, sound and energy reached a fairly high plateau of conceptualization. Julio Le Parc, Martha Boto and Heinz Mack were making light works. Le Parc, who was affiliated with the more scientifically oriented French Groupe de Recherche d'Art Visuel (G.R.A.V.), was engaged in hanging rows of small polished metal discs inside wooden boxes that were suspended from the ceiling. As the discs rotated, they glittered under the beams of a concealed light source (see *Ceiling Light Sculpture*, 1975 [no. 48]). Boto, affiliated with La Nouvelle Tendence, was making light works that were actual events on repetitive field structures as in *Variations Lumineuses, Paris*, 1965 (no. 11). Mack, a member of Group Zero in Germany, was working extensively with the phenomena of reflections, making monochromatic, sheet aluminum reliefs that resemble grid paintings in their structure. In *Light Dynamo No. 1*, 1966 (no. 49), light is "articulated and vibrated" from behind in shifting field patterns.

Jesus-Raphael Soto recognized the kinetic implications of line interference patterns, as revealed in his *Rond Blanc*, 1970 (no. 78), and also recognized and worked with the visual implications of field effects. Takis involved himself in investigations of magnetic fields as in his *Signal* series, 1968 (no. 82).

The Swiss-born Jean Tinguely, humorous and dystopian, is a pioneer of sound in art and also a pioneer of appropriation. Interested in the performance aspects of kinetics rather than the physics of kinetics, he combines machine technology with discarded industrial materials while quoting the concepts of Duchamp and the Dadaists. In witty assemblages and happenings Tinguely satirizes and disparages the dangers of life in a mechanized society with an emphasis on humanism rather than the programmatic. All of this is clearly demonstrated in *Tokyo Gal*, 1967 (no. 83), with its hilarious squawks and movements accompanied by sinister undertones.

Meanwhile in America, George Rickey, who was originally influenced by Calder and Gabo, was experimenting with kineticism. Subject to gravity and wind, the graceful *Tidal V*, 1963 (no. 68), and the brilliantly engineered and complex *Four Trapezoids as Two Rectangles II*, 1985-1987 (no. 69), reveal the power of Rickey's development and achievement. He has succeeded in fusing a constructivist heritage with technology, material, scale and innovation to create an organic aesthetic of motion.

Another first-rank kinetic artist was Len Lye. *Fountain*, (1959, no. 45), a manipulated spinning form, is reminiscent of the "virtual volumes" of Gabo and Moholy-Nagy with the important difference that he was able to realize, conceptually and mechanically, considerations which had only been outlined by those pioneers. It was Lye who produced the first successful type of motor-driven sculpture in America.

Until the mid-1960s, most technological experimentation in America had been conducted by isolated individuals, but a larger representation of these artists appeared on the scene soon after that. George Rhoads, originally a painter, began to make audiokinetic work in the 1960s. To this day the artist refuses to use electronics, preferring action that is entirely visible to the viewer. His spirited sculptures send balls propelled by gravity sliding down cylindrical, spiraling ramps. In turn, the balls set off bells, and the viewer who tracks its delightful visible mechanism is reminded of moving paintings and pinball machines.

Other artists in America who, like James Seawright, combine electronics with nature include Clyde Lynds, Alejandro and Moira Siña, and Rikuro Okamoto. In a public space, the lawn in front of downtown Santa Barbara's Lobero Theatre, Lynds has installed *Stele LXXV, Gnomon*, 1989 (no. 48), using fiber optics and concrete. Tiny holes in the concrete, which create a galaxy of slowly changing points of light created by the embedded fiber optics, call to mind starry constellations and the extraterrestrial. At the entrance of the Santa Barbara Museum of Natural History, the Siñas' *Night Butterfly*, 1980 (no. 74), greets evening visitors. A double fan of neon tubes connected to a swaying metal stem, the piece appears to dart and flutter as darkness sets in. A lyrical recreation of nature through technology, it differs in character from the artists' huge and powerful *Spinning Shaft*, 1984 (no. 75), with its whirling light bars that produce luminous spirals of color. Rikuro Okamoto makes "moving rocks," in this case *Moving Rock No. 22*, 1988 (no. 55). Critic Walter Robinson has described its movement as ". . . slower than a crab, back and forth across the gallery floor, with an effortless grace reminiscent of. . . the moon in a sky of scudding clouds."[5]

Milton Komisar's theatrical use of the computer in sculpture is high-art performance. On the technological side of things the artist constructs networks of plexiglas rods, joined together by faceted modules containing small light bulbs. Then he uses thin wires to connect these elements to a computer whose programming allows each rod to take on several hues in rapid, pulsating sequences. Visually, *The Philosopher's Geometry*, 1988 (no. 41), resembles linear abstraction and, subliminally, it suggests formally structured music. What makes Komisar's work so evocative is the disembodied, hypnotic effect of its delicate sequences.

Moving back and forth along its sleek, 16-foot-long brushed aluminum surface, punctuated by six tiny holes, viewers can activate and orchestrate Howard Jones' *Sonic Six*, 1967-1968 (no. 37). Without touching the artist's curved, carefully calibrated wall relief, the viewer's hand movements can control the pitch, tone, timbre, attack and delay of percussive sounds which emanate from Jones' sculpture.

François and Bernard Baschet match visual beauty with rich sound in works that are syntheses of shapes, sounds and public participation. In *Sculpture Musicale No. 20*, 1967 (no. 3), the sound of the wooden ball striking the vertical elements is magnified by the flaring metal trumpets of aluminum, which quiver visibly as the sound eddies through them. Delicate allusions to nature in *Fountain*, 1974 (no. 4), evoke botanical images and the sculpture chimes lightly each time a water cup strikes the tone bar.

Using simple, direct wiring, Jack Dollhausen is able to modify electronic circuitry into any configuration. The artist makes compositions of moving light, sound and space, which change and interact in step with the life around it, leaving his process of making as visible as possible. Viewers approaching *Too Close*, 1988 (no. 21) are reminded at first of an abstract drawing until interaction occurs with its playful, warning growl. In other works, for instance, *Midwest Summer Night's Dream*, 1977 (no. 20), Dollhausen makes romantic references to midsummer night dreams, night sounds and the flickering of fireflies.

A critical difference exists between these more familiar configurations of art and technology and some of the art made since the late 1960s by artists who do not make work about technology per se but rely on it as a means to examine other issues. In Nam June Paik's case, the issues revolve around manipulated video and happenings while Steven Cortright is engaged in critiquing the very media that Paik celebrates. Robert Irwin, Dan Flavin and Bruce Nauman use technology to support works of art that are concerned with phenomenology.

Robert Irwin has investigated perception for more than twenty years and his theories on the subject of light, space and perception have altered the definition of art itself. Essentially an optimistic modernist, his disc paintings (*Untitled*, 1969 [no. 35], is included in the exhibition but not reproduced here) in the late 1960s and early 1970s were successful in breaking the edge of the canvas with the result that the space around the canvas became equally important. His technological and philosophical experimentation in light, space and perception has evolved from object to non-object to site- conditioned/ determined sculpture. Irwin's work takes the position of placing responsibility upon the viewer in order to bring him to know the nature of his own presence.

No one else approaches the style in which Bruce Nauman has used neon. The majority of artists working with neon technology have used it in undeniably pretty forms or as decorative abstraction. Nauman, while investigating what sculpture is or might be, has used the technology of neon and the concept of advertising signage as one way to reject the ideals on which traditional sculpture was based and to reformulate another syntax for sculpture. *Malice*, 1980 (no. 54), involves letter inversion and a resonant word laden with a sense of vague, portentous dread. These colors, too, are pretty to look at and the proportions of *Malice* are beautifully formed. Nonetheless, it's a blunt, unmistakable symbol, anticipating the signboards of Jenny Holzer, which is explicitly involved in a phenomenological analysis of the viewer's relation to a cultural construct. Here color and form are merged with material, process, concept and irony, to make a compelling sculptural presence with a potential for social commentary.

Using cultural "icons" from the hardware store, Dan Flavin's austere fluorescent light constructions have a veracity and directness that emphasizes their mundane source. As Donald Judd pointed out long ago, the main aspects of Flavin's

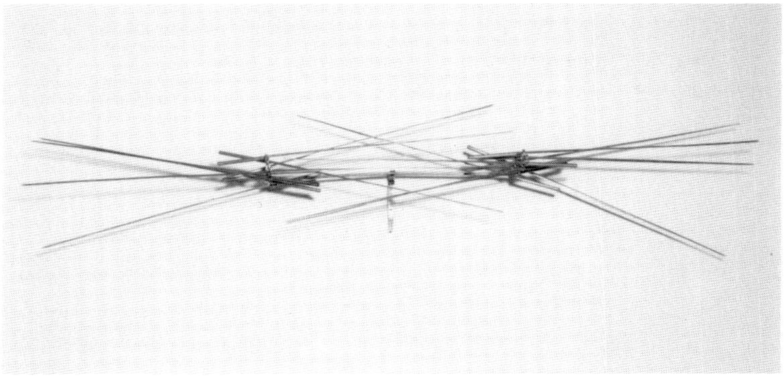

George Rickey
Tidal V
1963

work are the fluorescent tubes as the source of light, the light diffused throughout the surrounding space or cast upon nearby surfaces, and the arrangement together or placement upon surfaces of the fixtures and tubes. The lit tubes, intense and definite, are very much a particular visible state, a phenomenon.[6] Whether operating singly or in juxtaposition to one another in a light environment, their geometric purity can also become thoroughly theatrical in that Flavin's light sculptures respond to and rely on the spaces and surfaces that surround them for their context. Flavin's light can dissolve the surfaces of the floor and the walls and be the means by which all visual information about the space and any individual in it is revealed. The artist's increasing expertise and a clearly felt need to make light environments has, in recent years, made his work less direct and more theatrical. On the one hand, *Monument for V. Tatlin*, 1969 (no. 27), is involved with Flavin's pursuit of a particular phenomenon, but a Duchampian paradox occurs vis à vis *Untitled (Fondly, to Helen)*, 1976 (no. 28). Its new-wave colors can be interpreted as a critique on fluorescent technology itself or on an "arty" aura.

Bill Fontana's *Elevated Soundings*, 1990 (no. 29), a sculpture with no visual elements, is a sound process work concerned with ecological habitats. It offers listeners a cubist rendering of collaged and layered sounds that form an audio portrait of campus activities mixed with bird calls from distant places. For listeners, Fontana's sculpture brings about an intensified awareness of the shape of the landscape. The environmental conditions at any given moment determine how the sound is experienced.

A dada-like mix of sounds and images, *History of Animals* by Richard Dunlap, is to be performed at the University's College of Creative Studies in October. It is a sound and music performance within a sculptural context. Visual elements include modulating lights and projections, static and moving sculptural objects and original musical instruments. The sound/music events are both live and sequenced and use natural acoustics, electronically modified acoustics and digitally sampled sounds and sequences.

The visual images extend from pure spatial colorations to transformed projects of animals, creatures and man. Two screens (one using a scrim for front-back lighting effects) are used for projections and spatial ambience. Musical events include digital interfaces with one sound being transformed into another (multi-layered) and micro-tonally tuned pianos in tandem with a variety of animal sounds.

With a postmodern, low-tech sensibility that is akin to many of their European contemporaries, Bill and Mary Buchen make art with a casual aproach to materials. It's their exotic or amusing sounds and the viewer's interaction that are important to this artist-team. *Wind Gamelan*, 1987 (no. 15), takes the museum audience on a walk to nearby bluffs overlooking the Pacific. There the artists have erected an international composite of seven towers adapted from the forms of wind mills used on California wind farms. Each of the towers represents one tone in the Balinese and Javanese musical scale, and each chimes at intervals governed by wind currents that cause small hammers to strike tone bars on the towers. The Buchens' playful *Singer Songs*, 1980 (no. 14), is a modified treadle sewing machine equipped with a headset that plays recordings of folk and children's songs and machine noises.

Also at the University's College of Creative Studies, Mimi Garrard and James Seawright will present *Walking on Gravel*, 1990 (no. 30), a dance performance/light installation in October. Garrard will choreograph and Seawright will light a performance to include five dancers and a sculptural lighting installation integral to the theme of the work. Their collaboration permits light to be used as part of the process from the beginning, enabling Garrard to "use light to create and shape the movement."

The Survival Research Laboratories is a San Francisco-based collective formed in 1979 by Mark Pauline. The collective orchestrates elaborate performances with deafening sound levels. They perform in public places where people congregate rather than on art sites using industrial technology to illuminate man's experience in society. As simulations of war zones and street violence, these events are designed to force the viewer to consider some of the more shocking aspects of modern industrial culture and the politics of power. SRL's goal is to change the way viewers focus on political and social issues. Physically terrifying scenarios of anarchy are

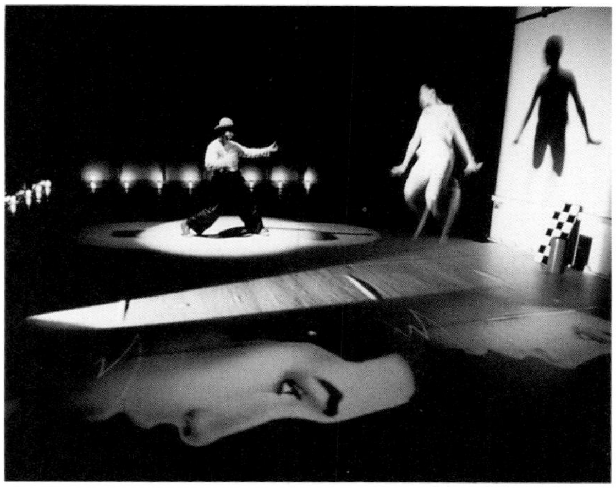

Mimi Garrard and James Seawright
Walking on Gravel
1990

produced by industrial cast-offs which are operated by remote control. Massive metallic behemoths attack each other and the simulated violence of the performance is not easy to accept. It breaks all the rules, it's not pretty, and it isn't contained in any way, shape or form by its materials, according to Philip Yenawine. Yet over and over again, SRL has received the full support of urban officials (police and firemen) in American cities who monitor these events in order to reassure the fascinated public. Their spectacles are controversial, striking a positive response with thousands of people but also striking a negative reaction among those who regard SRL's performances as cynical visions that trade in apocalyptic associations. A videotape of their *San Francisco Artspace Installation* will be available in the fall to museum visitors and Mark Pauline will speak at a public event on the campus in October.

With this kind of work, there seems to be no particular historical iconography based on the memory of previous work as there is in painting. Rather it is new twentieth-century attitudes and developing technologies that fuel the changing basis for its material, process, form and content. High-tech or low-tech, every artist included here is dealing with aspects of technology. Some of the messages emanating from the works in the exhibition could be perceived as pro-kinetic for by now kinetic art is an established fact of art history despite its failure to become the twentieth century's dominant form of artistic expression. But other sections cannot be mistaken for anything but compelling suggestions about individual and public concerns for today's conditional life style.

Art is in the process of being redefined as artists establish new concepts that deal with real existence and real time. Bearing in mind that art is a first-order phenomenon rather than a by-product and that it occurs in response to a set of specifics in our environment, it's not so much the art but some of the specifics of past decades that may appear dated right now. As those former states of awareness ceased to apply, the remote future arrived and art today is clearly connected to our changing environment.

These artists position their work in a fairly direct affiliation with contemporary questions, be it philosophical, perceptual, political or personal. Now more than ever before, they want their art to speak of universal experience and time, making it accessible to all. Increasingly, they believe that art must break through the confined atmospheres of museums and galleries, reaching out to the larger population. The future of art lies less with the creation of irreplaceable masterpieces and more with defining cultural strategies of adaptation.

A selection of old and new, high and low, known and unknown, this exhibition is not the last word on its subject, just the most recent.

Phyllis Plous is curator of the University Art Museum, University of California, Santa Barbara.

Footnotes

1. See Enrico Castellani's statement in Udo Kultermann's *Monochrome Malerei* (exhibition catalogue). Leverkusen: Städtisches Museum, 1960, 4-5.

2. George Rickey. "The New Tendency (Nouvelle Tendance Recherche Continuelle)." *Art Journal*, Summer 1964, 272-9.

3. Kenneth Baker. "Alan Rath's Video Rates Double Takes," *San Francisco Chronicle*, 15 May 1986, 68.

4. Germano Celant. "Rebecca Horn, Dancing on the Egg," *Artforum*, October 1984, 48-55.

5. See Walter Robinson, "Ricuro Okamoto at O.K. Harris," *Art in America*, December 1986, 132.

6. Don Judd. "Aspects of Flavin," *Arts and Artists*, March 1970, 48-9.

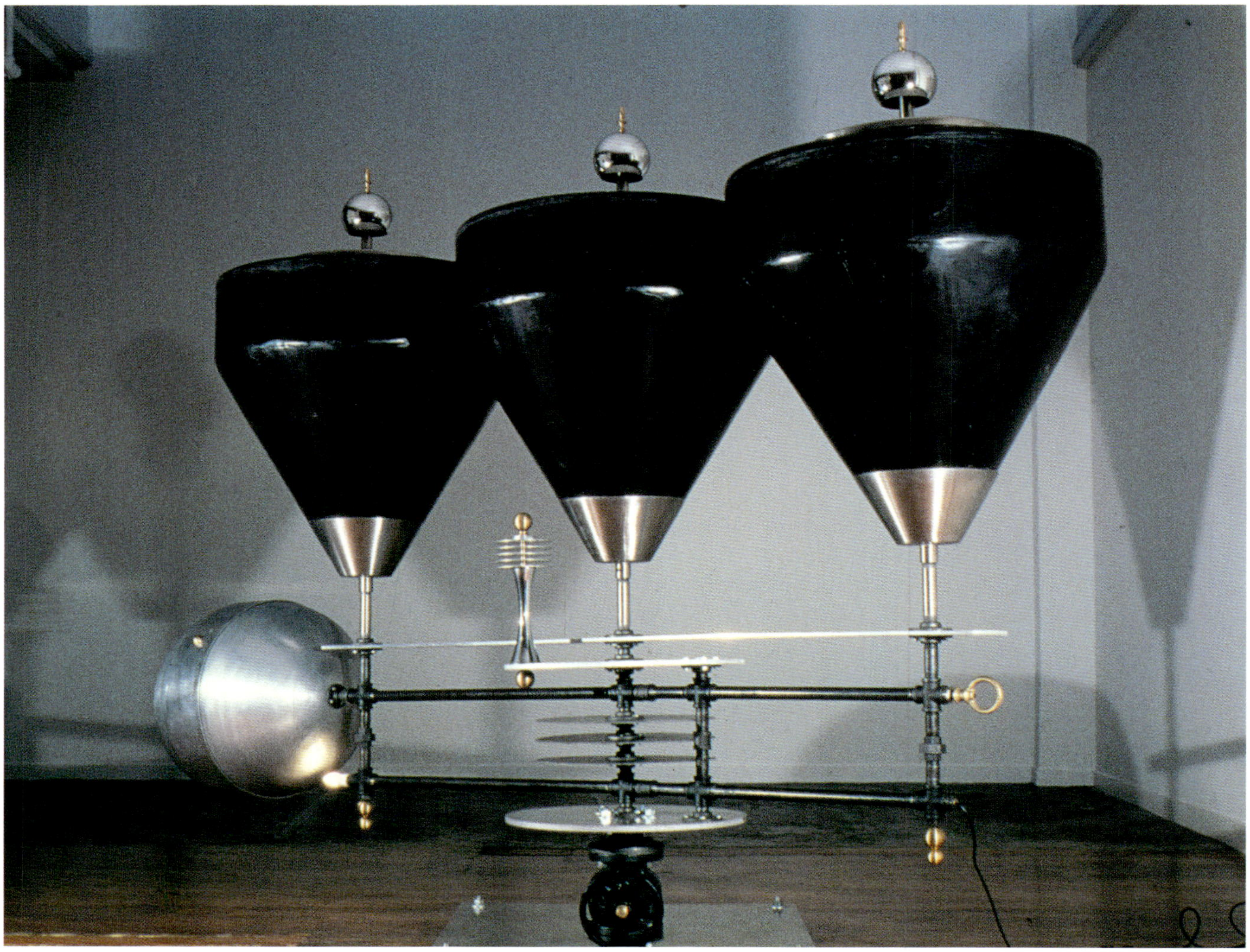

R. M. Fischer
Snap, Crackle, Pop
1986

Improbable Machines

Nancy Doll

> *The machine has become more than a mere adjunct of life. It is really a part of human life...perhaps the very soul. In seeking forms through which to interpret ideas or by which to expose human characteristics I have come at length upon the form which appears most brilliantly plastic and fraught with symbolism. I have enlisted the machinery of the modern world and introduced it into my studio.* —Francis Picabia, 1915

Artists have been making machines since well before the Industrial Revolution. The stage effects and automata of ancient Greece, the engineered visual experiments of the Renaissance, and the devices of Baroque spectacles all reflected an interaction between art and science. Although the relationship began to wane with the advent of the mechanical age in the early nineteenth century, the ubiquitous presence of machines in the fabric of modern life supports the idea of the machine as an element of culture as well as of science. And just as technology has played a major role in modern culture, through technology artists have introduced new forms and materials to art and further expanded its conceptual boundaries.

Artists in this century have regarded machines with a wide variety of interpretations and attitudes, ranging from distrust and pessimism to curiosity and romanticizing acclaim. The machine emerged as a prominent image in art in the early decades of the century, and such artists as Marcel Duchamp and Francis Picabia were forerunners in infusing it with symbolic meaning. Beyond the representation of machines (by the Futurists, Russian Constructivists, and Dadaists, among others), artists also began to incorporate real mechanisms into their work. As early as the onset of the 1920s, Duchamp, the pioneer of the modern artist-made machines, Naum Gabo, and László Moholy-Nagy all recognized the metaphoric potential of machines. Other artists such as Man Ray and Takis followed suit, making further contributions to the history of mechanized art in terms of both mechanical innovation and aesthetic expression. During the 1950s and 1960s artists continued to explore the expressive possibilities of motion in what came generally to be called "kinetic art"—work incorporating real movement in time. By the late 1960s, groups like Experiments in Art and Technology (E.A.T.) forged alliances between artists and engineers in an effort to conjoin scientific know-how with visual sensibility. Unfortunately, "art" was sometimes superseded by "science" in work that became technical gimmickry.

In the work of a number of artists today, the "gee whiz" aspect that characterized some earlier art machines has been replaced by a use of technology merged with intuition. These artists are not interested in simply demonstrating the capabilities of the latest devices. Instead, they are concerned with making art that operates on multiple levels and finds in the machine vocabulary a host of richly symbolic options.

Throughout his career, Dennis Oppenheim has been preoccupied with exploring the nature and activity of the creative process. This interest fueled the range of his conceptually oriented works in the later 1960s (including earthworks, body art, and performance art) and also characterized his return to making objects in the early 1970s; it remains central to his machineworks from the last dozen years. To Oppenheim, the workings of the machine run parallel to those of the human mind. His mechanical creations represent the visual equivalent of thought, of the circulation of ideas, and of the transmission of energy.

Though smaller and more intimate in nature, *Ghost Tantrums/Static Tremors,* 1988-1989 (no. 56), relates to the "factories" Oppenheim began in 1979—mammoth assemblages of interrelated parts that, like the mind, could run the spectrum from productive activity to "wheel-spinning" or even breakdown. In this case, a toy train engine repeatedly circuits a track supported by the cut-out letters of the work's title, passing distorting mirrors at each bend. The engine is crowned by a small welder's helmet, a symbol perhaps of the "ghost of the machine," the spirit of its maker, or the driving force of creative thought.

In a recent series of "tool pieces," Oppenheim has brought the making of a work into the final product, as if to make tangible the conceptual energy that stimulated it. *Four Spinning Dancers* (from *The Power Tools Series*), 1989 (no. 57), poses a group of whirling figures made of stacked buffing disks on a shelf in front of a reflecting panel; the dolls are driven by individual electric drills. In such works, Oppenheim deftly balances concept, form, and function with a sense of play and a knack for imbuing objects with a heightened reality.

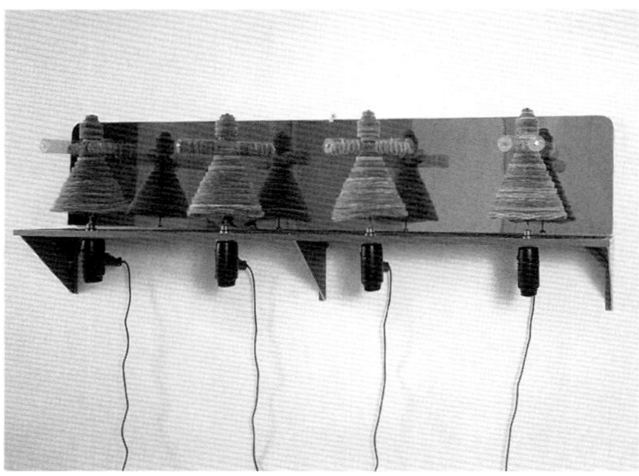

Dennis Oppenheim
Four Spinning Dancers (from *The Power Tools Series*)
1989

The spirit of inquiry and metaphoric possibility underlying Oppenheim's work is also at the heart of Alice Aycock's enigmatic mechanized sculptures. Her works have been described as exploring the little-charted territory "where science and magic meet." Among their many strengths—which include beautiful craftsmanship and elegant movement—is the artist's ability to layer them with a full panoply of historical, scientific, allegorical, and personal references. Thus richly imbued, they serve as symbols for humanity's quest for meaning and for the artist's explorations beyond mere form and aesthetics.

Following a natural sequence, Aycock moved from works based on vernacular building forms that solicited movement by the viewer through them to works that were physically activated themselves, usually by motors. In 1982, she began a prolific series of spinning blade sculptures called *Time/Creation Machines*, the designs for which were based on medieval and renaissance alchemical drawings about the creation of matter. Linking modern technology with the historical and mythical past, these works appear to have a distinct, if elusive, function having something to do with undetected or inexplicable phenomena in the universe.

Fata Morgana (Of Things Seen in the Sky), 1984 (no. 1), combines spinning blades, an arc of orange neon, and a grinding wheel in a floor apparatus that is reflected in a mirror structure suspended from the ceiling. Both the title and the form of the piece refer to Morgan Le Fay, sister of King Arthur and sorceress, who could conjure floating apparitions. Like Le Fay, the artist, too, is a kind of conjurer, captivating the viewer and summoning an awareness that extends beyond simple sensory perception. In art, as in alchemy, Aycock sees knowledge and imagination operating hand in hand.

Although viewers are challenged to look beyond the mere physical form and technical process of these machines for multiple and complex levels of content, they may also be asked to look into the mechanical heart of the work. In an age of electronics, where so much is ephemeral and invisible, artists such as Jon Kessler and Roland Brener expose the "guts" of their works, thereby demystifying them and rendering them approachable.

Throughout much of the past decade, Jon Kessler was known for his kitsch-filled tableaux that freely associated all manner of cultural debris and projected a mood of nostalgia and quiet humor. More recently, he has chosen to work with sparer forms, emphasizing their qualities of light, structure, and motion. Despite their cool technology, these works achieve a lyrical and sympathetic sense. A further paradox is the fact that, in an almost modernist observance of the "truth to materials" dictum, their mechanisms are exposed and their inner works revealed, yet they retain an aura of magic.

Some of Kessler's recent works quote architectural forms with a certain elegiac tone. The freestanding *Blackout*, 1989 (no. 40), combines frosted glass, aluminum, and steel mesh in a tubular form that suggests a modernist skyscraper. The motorized light making its way up and down a channel in the form alternately attracts us and keeps us at a distance as it shuts on and off. *16 Rue Chanoinesse*, 1989 (no. 39), merges past and present in its combination of technology with a more traditionally planar photographic image of a spiral staircase. In motion, the revolving image is mesmerizing; it is also a sly reference to Duchamp's *Rotoreliefs*, the progenitor of the twentieth-century artists' machines.

The work of R.M. Fischer shares with Kessler's an eclectic combination of objects and materials from our industrial and consumerist culture, as well as a sense of generally understated humor and nostalgic reverie. Since the late 1970s, Fischer's lamp works have posed a dual reading: as abstract, assembled sculptures and as real-world, functional objects. At once art object and tool, each piece strikes a different balance between elegance and ungainliness, although the scales always tip toward refined grace. The works subvert the world of design and technology from which they derive; their world is both antique and forward-looking, but they seem more akin to those futuristic designs visualized in the 1930s than to the images of space-era technology. Fischer's lamp sculptures are from the age of hardware rather than electronics, from a time when machines often seemed anthropomorphic, even "user-friendly" before the term was coined.

Ambiguity, including a sense of the works existing in more than one world, is central to Fischer's art. His works have always, in a sense, re-invented themselves, moving from the real world of functional objects, into the realm of art, and back again. The earlier lamps were often assemblages of found and cast-off materials. More recently, Fischer has turned to manufactured hardware and industrial elements, modifying them or altering their context. The resulting works are more refined in appearance and also more sculptural, connoting less pointedly the idea of lamp or furniture.

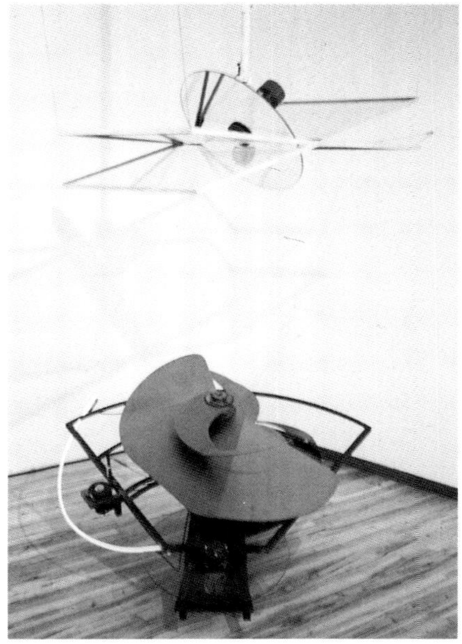

Alice Aycock
Fata Morgana (Of Things Seen in the Sky)
1984

Snap, Crackle, Pop, 1986 (no. 25), suggests both a triad of elements from a power plant and some type of farm equipment. This shift between a potentially ominous and a more positive reading is underscored by the work's ironic title. *Amadeus*, 1986 (no. 26), looking like a baroque record player, implies movement without actually incorporating motion. Odd and unlikely as these works may be, they are also refreshingly accessible in an age of circuitry incomprehensible to most. Plugged simply into the wall and warmed by the glow of their electric lights, Fischer's works bespeak an abiding faith in the visual ideals of the modernist era.

The work of Roland Brener also seems to participate in several worlds at once. Operating on the cusp between art and society, it is situated within both the tradition of sculptural objects and the context of contemporary popular culture. Like Kessler and Fischer, Brener favors using the everyday things around us, often culled from the world of consumer objects, which offer a sense of familiarity and accessibility to the viewer. The artist relies upon this empathic relationship as the basis from which he can alter the form, function, or context of his chosen materials, giving them new and unexpected meaning.

The central image of *Small Talk*, 1987 (no. 12), is the head of a popular children's talking toy, "Teddy Ruxpin," which has been stripped of its fur and stuffing to reveal its working mechanism and placed behind a magnifying glass. The character of this teddy bear is thus transformed from benign to somewhat ominous. As in many of Brener's works, the viewer's presence activates the piece, setting off an incomprehensible strain of "talk" from Teddy's mouth. The suggestion of meaningless communication and failed desire is poignant, and the piece exemplifies the way Brener's assemblages point to the dichotomy of faith and mistrust we feel toward machines.

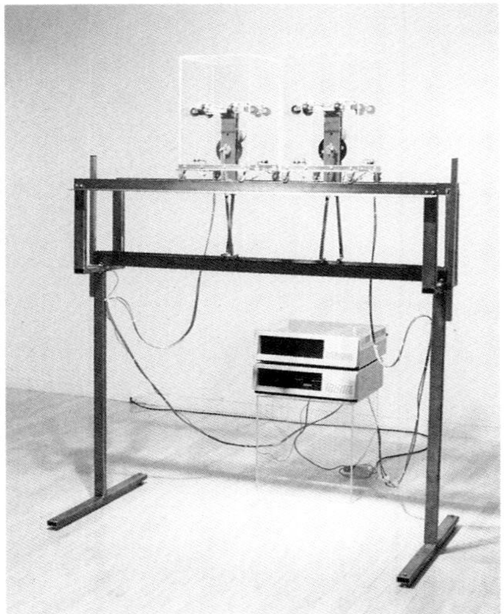

Roland Brener
Life in Quick Basic
1987

In *Life in Quick Basic*, 1987-1988 (no. 13), two primitive mechanical robots move across an elevated rail through a series of seven "routines" that mimic human behavior—at times moving independently, at times in response to one another—to the background hum of driving motors. The title of the piece refers to the animated activity of the robots and to the computer language with which they are programmed. Again, the engaging nature of the piece is countered by a slightly dire reference to our increasingly electronic future.

In contrast to the feelings of mistrust or intimidation with which we sometimes confront machines, some artists, including Steve Barry, present a sympathetic response. Barry's machines remain machines, but he endows them with a humane—if not human—character; and viewers are engaged aesthetically, emotionally, psychologically, and, usually, physically. For the last several years, Barry has invited, and depended upon, the viewer's involvement in activating what have been called his "mechanical performance" installations. To Barry, this is neither a gimmick nor an exercise in technical know-how, but rather a means to restructuring the passive way in which we routinely perceive art objects. Barry's work confronts us, involves us as accomplices, and sometimes places us in a disquieting situation.

A recent series of provocative installation pieces named after figures from classical Greek myths combine industrial-looking forms with allegorical references. These works pose a dichotomy between large scale and intimate experience. Similarly, *I.D.*, 1989 (no. 2), is a self-contained mechanical device—a bulky form built primarily from scrap materials—that offers a poetic and personal moment to the viewer. *I.D.* departs from the common experience of having one's picture taken in a photo booth, perhaps "for posterity." Inside Barry's booth, the viewer sits facing a mirror and in front of a landscape photograph; when the machine is activated, the viewer's image disappears into a reflected view of nature. This experience recalls the belief of certain primitive peoples that the camera robs one's soul; to Barry, however, it is less a pessimistic reverie on the inevitability of loss or death than an affirmation of the present. Even if momentarily, the viewer is encouraged toward self-awareness, realizing how much remains when one's image disappears.

The mechanized creations of all of these artists retain their character as machines—they *do* function, they *do* convey energy—but theirs is a nature quite beyond the conventionally mechanical. In their emphasis on imagination, these machine-works are individual and unique rather than typically reproducible; moreover, whereas conventional machines strive toward operational perfection, these artists' machines are imperfect and improbable. As recipients of technology, these artists pose a connection with the rest of us who use technology, accepting it as part of our daily reality but rarely giving it much consideration. In the hands of these artists, machines are liberated from the rules and systems usually governing them by the freer values of artistic thought.

Nancy Doll is curator of twentieth century art at the Santa Barbara Museum of Art.

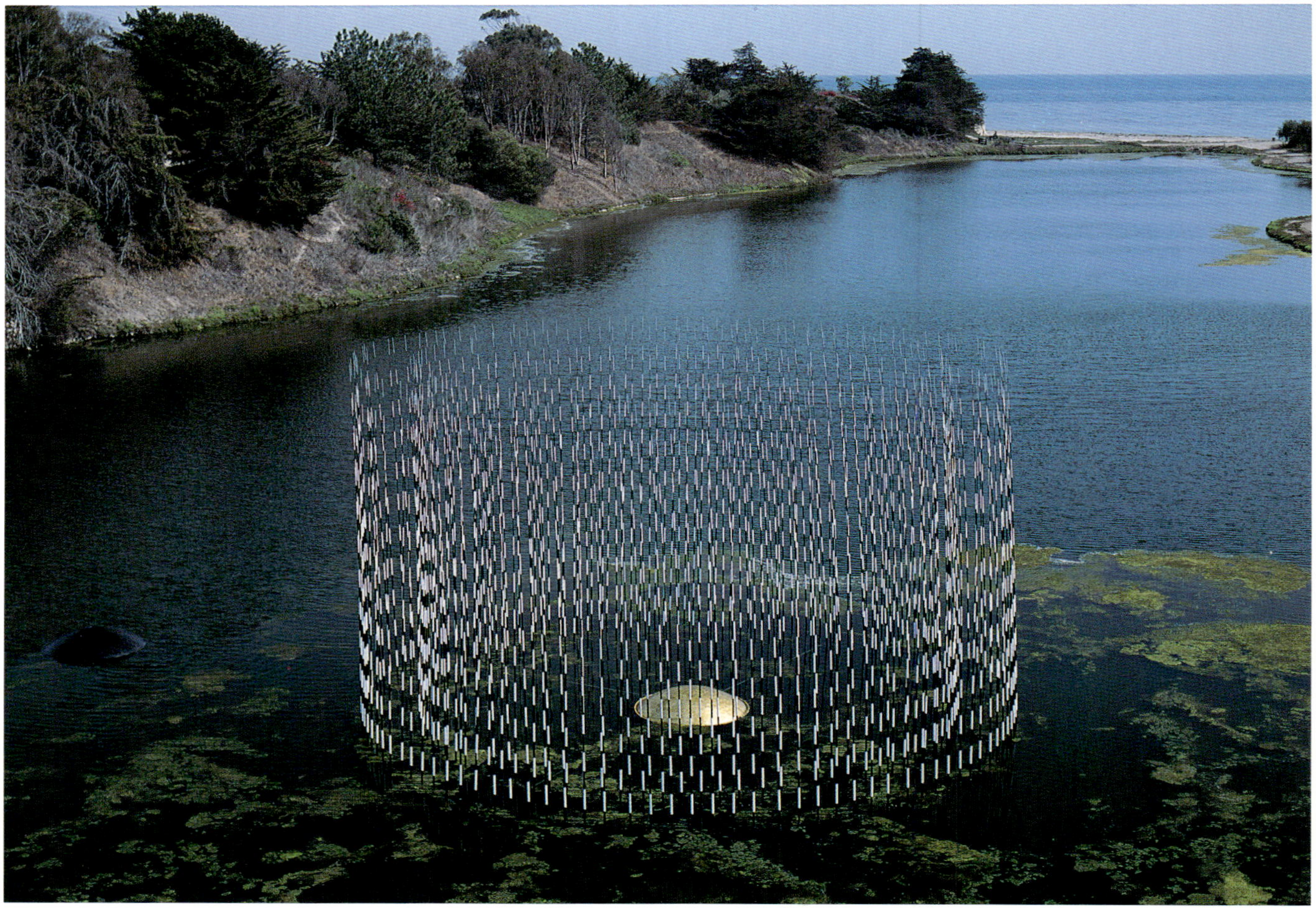

Kristin Jones and Andrew Ginzel
Pacific Lagoon Project
1990

Sounds and Spectacles:
Artists Working with Developing Technologies

Diane Shamash

Like acts of nature (a steaming volcano, an earthquake, lightning) technology often evokes in people feelings of excitement, fear, suspense, dread or imminent danger. Artists have long made use of new technologies because of their associations with utopia and the apocalypse, and also as a means of disseminating their work.

In the early part of the 20th century, artists such as Hungarian-born Laszlo Moholy-Nagy and Russian Constructivists Tatlin, El Lissitzky and Rodchenko experimented with new technologies to create interdisciplinary works in print, sculpture, architecture, music, theater, and performance. For these artists and others, technology was a means to reinvent the language of art and redirect its attention to a radical agenda of social reconstruction in Russia and Eastern Europe. Rapidly assimilating new media and distribution methods into their work, these artists moved their artworks into the streets, cafes, and bookstores of Moscow, Leningrad, Berlin, and later to the cities of the United States.

By the 1930s a growing number of artists and artist collectives in the United States, most notably the New Bauhaus in Chicago, were dedicated to redefining the form of and audience for art. Directed by Moholy-Nagy, the New Bauhaus was created in 1937 "to form a nucleus for an independent, reliable center where art, science and technology will be united in a creative pattern." [1]

Artist Gyorgy Kepes directed the light and color department at the New Bauhaus, and later founded the Center for Advanced Visual Studies (CAVS) at MIT to provide a laboratory where artists and scientists could collaborate on projects experimenting with new media. During his tenure at MIT from 1945-1977, Kepes collaborated with architects and planners such as Gropius and Kevin Lynch and co-directed the study *Perceptual Forms of the City* with Kevin Lynch from 1954-1958. This important study emphasized the need to provide a visual and perceptual understanding of city form, a need that was later articulated in Lynch's influential book *The Image of the City*.

Both Kepes and Moholy-Nagy were visionaries and educators committed to integrating art into the mainstream of daily life. They were vocal advocates of an art connected to the world at large—for artists working collaboratively with architects and city planners, engineers, scientists, and other individuals of imagination who would change the visual and social structure of society. Kepes stated: "Artists have come to recognize that their creative imagination and sensibilities are neither self-generated nor self-contained; they belong to a larger environmental field of nature and society." [2]

If these technologically driven art activities in the United States, Europe, and Russia in the 1920s and 1930s revealed a utopian or optimistic vision of society, they were also concerned with the specter of natural and human destruction made possible by this new technology. "What we now face are destructive forces of a completely new kind—man generated, cumulative, and of almost cosmic proportion..." [3] Like their colleagues in the sciences, artists became increasingly aware of technology's connection to mortality, survival and global interdependence.

By the 1950s a proliferation of artists and collectives that embraced new media with divergent aesthetic agendas emerged internationally.[4] Due to a lack of acceptance by the mainstream art world of so-called "technological art" (and a lack of interest in traditional venues on the part of artists themselves) these artists' works in light, sound, video, and computers often found alternative public audiences away from the gallery, museum, or concert hall.

Artist Nam June Paik's experimental works in video were part of FLUXUS "happenings" in the 1960s which took the form of guerrilla theater, street spectacles and concerts of electronic music; Sonia Sheridan's radical work with color xerography and computers earned her a place as the first artist-in-residence at 3M Corporation; Otto Piene, current director of CAVS, brought his multimedia light "ballets" to public audiences in the United States and Europe; and artists Robert Irwin and James Turrell conducted perceptual research at Garrett Corporation in 1969. These were only a few of the projects that helped redefine how art was understood.

Speaking about their collaboration at Garrett which was sponsored by the Los Angeles County Museum of Art's *Art and Technology Program*, Irwin says, "A problem may arise with this project in the minds of the art community, who may regard it as 'non-art'— as theatrical, or more scientific than artistic, or as being outside the arena of art. Although there is a strong alteration as far as method, means, and intent, we believe in it as art, and yet recognize the possibility of a redefinition needed to incorporate it into the 'arena'." [5]

What Irwin and other contemporary artists had in common with Moholy-Nagy and Kepes was their willingness to give up painting (and all its historic associations) in order to respond to the world at large—to its perceptual and social environment, and to bring to it any means and media deemed appropriate.

There is possibly no better place in 1990 to examine the issues raised by Robert Irwin and artists working with new technologies than the *PULSE 2* exhibition. The eighty-seven artworks by sixty artists that are represented in this exhibition involve sound, light, and energy and range from electronic sound sculpture to interactive video installations.

Although the *PULSE 2* exhibition brings together widely disparate artworks under the umbrella of art and technology, it clearly represents how artists using new technologies have redefined how we think about art and its public. It brings to the campus's lagoon, wooded walks and meeting places public artworks that resensitize people to the sounds and experiences of the campus, and that call into question how science and technology present the dual possibilities of paradise and destruction.

Installations and performances by Seattle composer David Mahler and San Francisco artist Bill Fontana ask us to tune into the natural and social sounds of the campus by creating interactive works that combine live and recorded sound. Bill Fontana's impressive *Elevated Soundings,* 1990 (no. 29), involves sonic projections at the Storke Tower on the campus and at the County Courthouse in downtown Santa Barbara.

As you walk through the campus, unaware of two weatherproof speakers mounted on the top of Storke Tower, you hear clear and penetrating bird calls such as sea gulls,

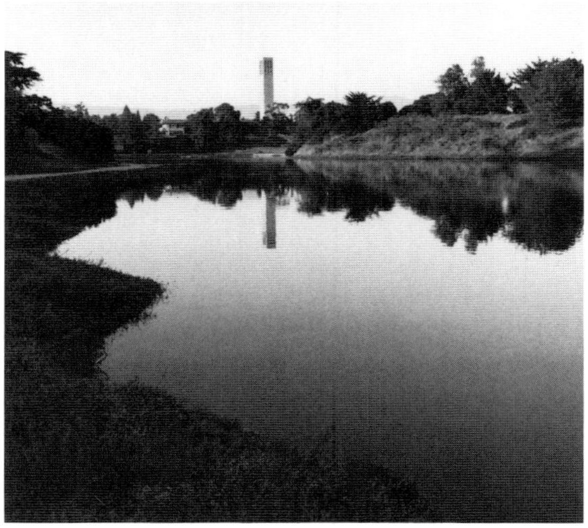

Storke Bell Tower
University of California
Santa Barbara

mockingbirds, nightingales, ravens and Japanese larks. As you approach Storke Tower, you become aware of ambient traffic sounds and live recordings from pedestrian pathways and tennis courts (e.g., the rhythmic bouncing of a ball at regular intervals).

These sounds, which tend to sharpen your listening skills, are punctuated by the Storke Tower bell sounding at ten minutes to the hour and then again exactly on the hour. Audible first in the seating area and later at the tower, these loud and ceremonial bells ring prominently in the landscape against the backdrop of bird songs and tennis balls.

There's no doubt that Bill Fontana's project intensifies the sound experience of the campus and raises an awareness of the campus as an environment that splices together natural and man-made habitats. By mixing live natural and human sounds (bells, planes, balls bouncing, people talking) with recorded ones (bird calls from exotic places), Fontana asks us to think about the connection between sound and habitat, music and noise, and the landscape that results: "The concept of ecology is used to describe the harmonious relationships existing between living species in natural habitats that enable them to mutually survive together. In these natural habitats, ecology can also be understood as being successful design relationships between the various aspects of environment." [6]

David Mahler, a Seattle composer and performer who creates works for audio tape, electronics, voice, instruments and ensembles, is also interested in sound observation. For the *PULSE 2* exhibition, Mahler will create an "aural portrait of the campus" using live and recorded sound from a variety of sources. Spending seven days on campus to "catch" sounds such as laughter and conversation, Mahler will create a sound map of the campus by editing and mixing these recorded sounds with live performance. His performance is scheduled for October 1990.

Speaking about the value of an audio tape recorder as a musical instrument and a method of sound observation, Mahler states: "I feel there is nothing that has happened in the past 50 years regarding music that is nearly as important as the introduction of the magnetic tape recorder. It is to sound what the camera is to sight...moreover, the medium for listening has now shifted from the church or drawing room or parlor or concert hall or bar to the audio speaker, which of course can be played in any location...through tape, sound has become portable." [7]

With a remarkably eclectic musical background that includes involvement with church choirs and marching bands as well as new electronics, Mahler's performances are often a

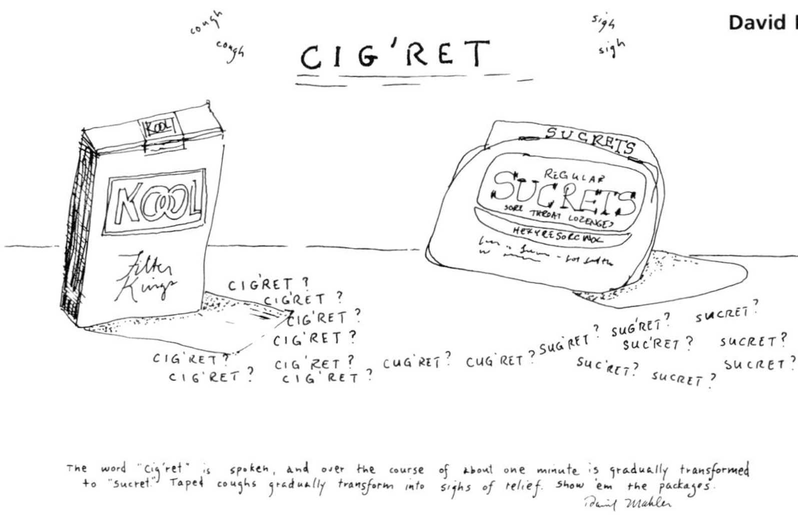

David Mahler
sketch
1990

mix of recorded and live audio recordings, ragtime piano, performance art and storytelling. His public projects, including *Listener in Residence & The Art of Tape,* 1990 (no. 50), in the exhibition, are about the enjoyment of listening and remembering a sound's relationship to its social setting.

Liz Phillips' installations also invite viewers to listen—in her case, to sounds created by an electronic system using computers, amplifiers, and speakers that channel, analyze, and amplify sounds in response to the viewer. People's movements are detected by capacitance fields that electronically sense the nearness of people to the objects that radiate them. The resulting change in the timbre, pitch, and volume of sound in the space dramatically alters the viewers' perception of the space and makes them wonder, "How does it work?, "How do my actions influence the sounds?" "What is this?"

Soundtable I and II (Wet and Dry Landscape) consists of two lacquered black tables and two speakers in opposite corners of a room that is barely lit. One table contains a dry rock garden made of natural copper ore formations and raw combed wool (shards of copper conduct and radiate capacitance fields); the second tabletop is filled with water and contains a small copper ribbon. As people move near the tables, their presence activates sounds that do not repeat

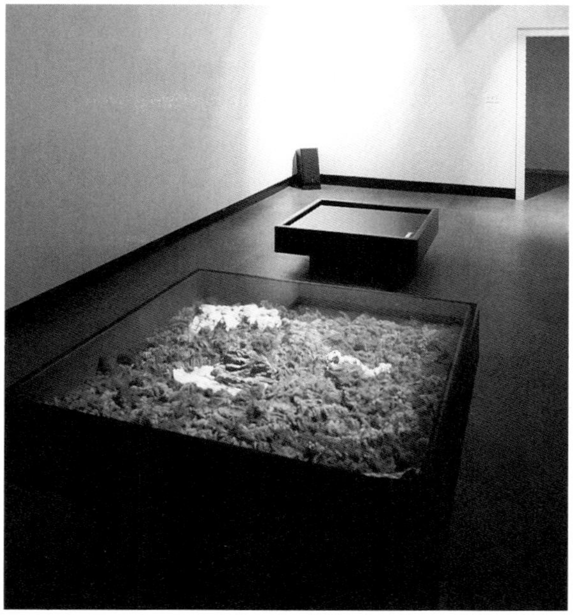

Liz Phillips
Soundtable I and II
(Wet and Dry Landscape)
1989-1990

themselves but shift and echo at changing pitches and volumes. Despite the clearly visible computer systems housed underneath the tables, the inner workings of this installation are entirely mysterious.

How technology defines our relationship to the world is evoked in the dense layering of images and sounds in this artwork. Sheep's wool, lava rock, and copper formations are contained in almost clinically scientific display boxes. Bursts of synthesized sound break into the stillness of the room. A miniature rock garden and raw sheep's wool obscure an Apple IIe computer and synthesizer underneath. The result is a demanding installation that disorients you in much the way a trip on *Challenger II* might. It requires a realignment of your body's orientation to physical space and sound, and encourages you to fantasize about primordial beginnings and unidentifiable objects and experiences.

Like Liz Phillips, Kristin Jones and Andrew Ginzel are interested in technology's ability to heighten our experiences in the world, as well as present its destructive potential. A central metaphor of their ephemeral, illusionistic, and highly theatrical artworks is the relationship of nature to human mortality. Often building the equivalent of a proscenium stage in a museum or public space, Jones and Ginzel mechanically animate elaborate still lifes that are more spectacles than objects.

"We seek to magnify the physics of existence within the contemporary urban context, in which the forces of nature are seemingly extinct....We are primarily interested in the phenomena of life itself, in the fact that this planet and universe are in flux, in the fact that we are alive at this instant." [8]

For the *Pacific Lagoon Project,* 1990 (no. 38), at the UCSB campus, Jones and Ginzel have built a 100-foot circular ring of black and white fiberglass tapers that emerge from the water in the northwest corner of the lagoon. Appearing to be a curious fortress or island, this tall cluster of poles responds like an instrument or stand of reeds to unseen events both above and below the water. Inside the circle of poles is a gold leaf dome that reflects and refracts changing levels of light.

Selecting a lagoon as the backdrop for the *Pacific Lagoon Project* represents these artists' continued interest in working with all forms of water: streams, whirlpools, boiling vessels, steam, fog, illuminated comets of water, rising and falling tides, effervescence. Presenting the physical and metaphoric properties of water as a reflective liquid surface and image, Ginzel and Jones seek to create "a striking apparition with a life of its own." [9]

Like their other multimedia artworks, this project offers striking geometry and metaphorical images—this time animated by actual natural phenomena (light, rain, dew, darkness, etc.). Although the motors and pumps of their gallery and museum installation are gone, Jones and Ginzel have once again created a magic show about the universe which urges us to use our senses to fantasize about real and imaginary forces.

If both the Liz Phillips and Jones and Ginzel projects explore technology's expressive and experiential potential, Michael Hardesty and Jenny Holzer examine its political and social consequence. In his indoor installation *As If,* 1990 (no. 31), the viewer approaches a large iridescent ball evocative of a globe, earth, sun, or solar system, which is 48 feet in diameter. Above this glowing ball is a commercially made LED (light-emitting diode) sign that repetitively displays the lines:

> SULFUR DIOXIDE CARBON WATER VAPOR
> HYDROGEN COMPOUNDS NITROGEN OXYGEN
> HELIUM ARGON NEON [10]

Clearly presenting a dim view of the destructive powers of technology, Hardesty transforms a small gallery room at the University Museum into a sort of observation chamber. In this chamber, the people observe the sculpture, which moves methodically and without response to viewers, the people observe each other, and then they leave the room.

Around the corner in the Museum is Jenny Holzer's *More Survival* (no. 33), produced in 1985. Since Jenny Holzer first began publishing her written work in 1977, anonymously, her *Truisms* and *Living and Survival* works have appeared in books, posters, billboards, LED signs and T-shirts among other venues.

While *Truisms* included a series of statements that were alphabetically ordered and that expressed seemingly incompatible points of views and opinions, the *Living and Survival* works were written in a personal and confrontational tone, using official formats such as bronze and aluminum plaques, silver stickers, and computerized signs.

In *More Survival*, statements are displayed on a computer-driven LED sign with red diodes and 2-inch characters. Like her other LED works, the methodical pace of the computer establishes an authoritative rhythm that both rivets and detaches the viewer from the LED's contradictory statements and opinions. Holzer's work both questions the way mass media and new technologies give and control access to information, and how the private voice can remain uncoopted by it. Most importantly, it seeks to make visible the gender, class, and special interests of what we tend to think of as a neutral technology.

The connecting threads between human beings and technology are also the subject of several anthropomorphic and humanoid artworks by Jonathan Borofsky and Rebecca Horn. Borofsky's *Eggs, Projector No. 3,214,872,* and *No. 3,214,875* include fiberglass eggs lying in the corner of the gallery next to two fiberglass women and a child all with holes in their foreheads to emit projected light. Installed together they create a futuristic scene that is reminiscent of 1950s films like *Invasion of the Body Snatchers*.

The sculpted figures seem frozen in place, as if photographically, and the sophisticated fiberglass material is made to look like cement. By combining low and high-tech materials, and by linking the image of a human figure with that of a machine or motor, Borofsky poses the question of how our physical and social bodies are driven and shaped by machines. Rebecca Horn's *Butterfly Machine* also combines the image of a living being with that of a machine. Parrot wings, which are soldered to a mechanical box, slowly open and close and click and whir like a metronome.

The exhibition brings together three generations of artists who have utilized technology in their work, from Lászlo Moholy-Nagy's *Light Display – Black and White and Gray* to Jenny Holzer's LED artwork, *More Survival*. Whether utopian or apocalyptic in their world view, artists in the 1920s and 1990s have constantly assimilated new media into their work in order to redefine our relationship to the world at large. Like scientists, they observe, identify, describe, investigate, and explain natural phenomena. It's time that their work was understood not as an art style, but as an ongoing investigation inseparable from life.

Diane Shamash is program coordinator for the Seattle Arts Commission.

Footnotes

1. László Moholy-Nagy. Quoted in Marjorie Supovitz, *Gyorgy Kepes: MIT Years 1945 - 1977* (Cambridge, MA.: The MIT Press, 1978), 7.

2. Gyorgy Kepes. Quoted in Marjorie Supovitz, *Gyorgy Kepes: MIT Years 1945 - 1977* (Cambridge, MA.: The MIT Press, 1978), 5.

3. *ibid.*, p.5.

4. Many artist collectives in the 1950s and 1960s emerged such as Experiments in Art and Technology (E.A.T.), founded by Billy Kluver in 1966 in New York City involving artists John Cage, Buckminster Fuller, Gyorgy Kepes and others; and the Zero Group (Zero refers to zone of silence from which new sounds would emerge) was founded in 1957 in Düsseldorf by Otto Piene and Heinz Mack.

5. Robert Irwin. Quoted in Maurice Tuchman, *Art and Technology,* New York: The Viking Press, Inc., 1971, 131.

6. Bill Fontana. Quoted in exhibition catalog, *Landscape Soundings* Wein: Wiener Festwochen, 1990, 6.

7. David Mahler. Quoted in publication, *A Celebration of Audio Tape* (Seattle: David Mahler), 2.

8. Kristin Jones and Andrew Ginzel. Quoted in exhibition catalog, *Kristin Jones and Andrew Ginzel.* Basel, Switzerland: Kunsthalle, Basel, 1989, 15.

9. Kristin Jones and Andrew Ginzel. Quoted in July 25th letter to Diane Shamash.

10. Michael Hardesty. Text from installation *As If*, University Art Museum, Santa Barbara, July - October 1990.

PULSE•2

Alice AYCOCK Steve BARRY François and Bernard BASCHET Bill BELL Harry BERTOIA Jonathan BOROFSKY Martha BOTO Roland BRENER Bill and Mary BUCHEN Alexander CALDER James O. CLARK Steven CORTRIGHT Jack DOLLHAUSEN Marcel DUCHAMP Richard DUNLAP R. M. FISCHER Dan FLAVIN Bill FONTANA Mimi GARRARD and James SEAWRIGHT Michael HARDESTY Michael HAYDEN Jenny HOLZER Rebecca HORN Robert IRWIN Joe JONES Howard JONES Kristin JONES and Andrew GINZEL Jon KESSLER Milton KOMISAR Harold LEHR Julio LE PARC Evan LEWIS Len LYE Clyde LYNDS Heinz MACK David MAHLER Ron MALLORY László MOHOLY-NAGY Bruce NAUMAN Rikuro OKAMOTO Dennis OPPENHEIM Eric ORR Nam June PAIK Liz PHILLIPS Alan RATH George RHOADS George RICKEY Raúl RUIZ James SEAWRIGHT Alejandro and Moira SIÑA Jesus-Raphael SOTO Art SPELLINGS SURVIVAL RESEARCH LABORATORIES Vasilli TAKIS Jean TINGUELY Ted VICTORIA Peter VOGEL Alice AYCOCK Steve BARRY François and Bernard BASCHET Bill BELL Harry BERTOIA Jonathan BOROFSKY Martha BOTO Roland BRENER Bill and Mary BUCHEN Alexander CALDER James O. CLARK Steven CORTRIGHT Jack DOLLHAUSEN Marcel DUCHAMP Richard DUNLAP R. M. FISCHER Dan FLAVIN Bill FONTANA Mimi GARRARD and James SEAWRIGHT Michael HARDESTY Michael HAYDEN Jenny HOLZER Rebecca HORN Robert IRWIN Joe JONES Howard JONES Kristin JONES and Andrew GINZEL Jon KESSLER Milton KOMISAR Harold LEHR Julio LE PARC Evan LEWIS Len LYE Clyde LYNDS Heinz MACK David MAHLER Ron MALLORY László MOHOLY-NAGY Bruce NAUMAN Rikuro OKAMOTO Dennis OPPENHEIM Eric ORR Nam June PAIK Liz PHILLIPS Alan RATH George RHOADS George RICKEY Raúl RUIZ James SEAWRIGHT Alejandro and Moira SIÑA Jesus-Raphael SOTO Art SPELLINGS SURVIVAL RESEARCH LABORATORIES Vasilli TAKIS Jean TINGUELY Ted VICTORIA Peter VOGEL Alice AYCOCK Steve BARRY François and Bernard BASCHET Bill BELL Harry BERTOIA Jonathan BOROFSKY Martha BOTO Roland BRENER Bill

Bruce Nauman
Malice
1980

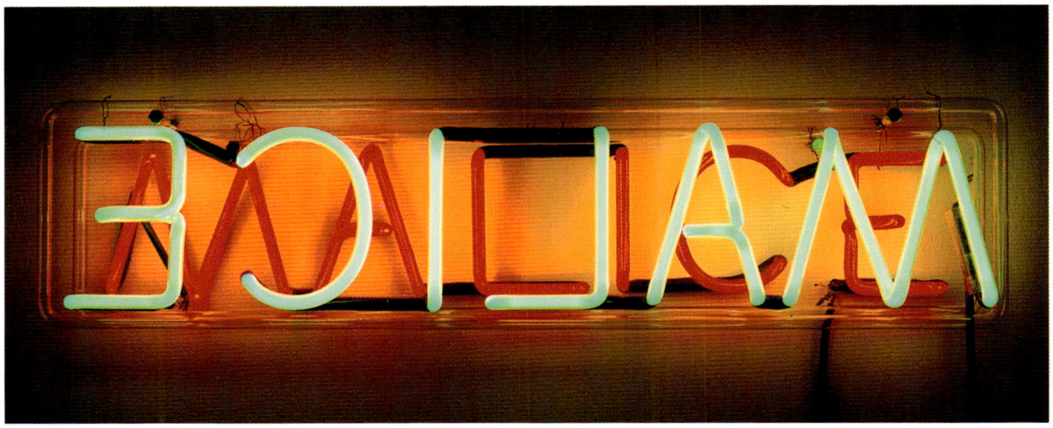

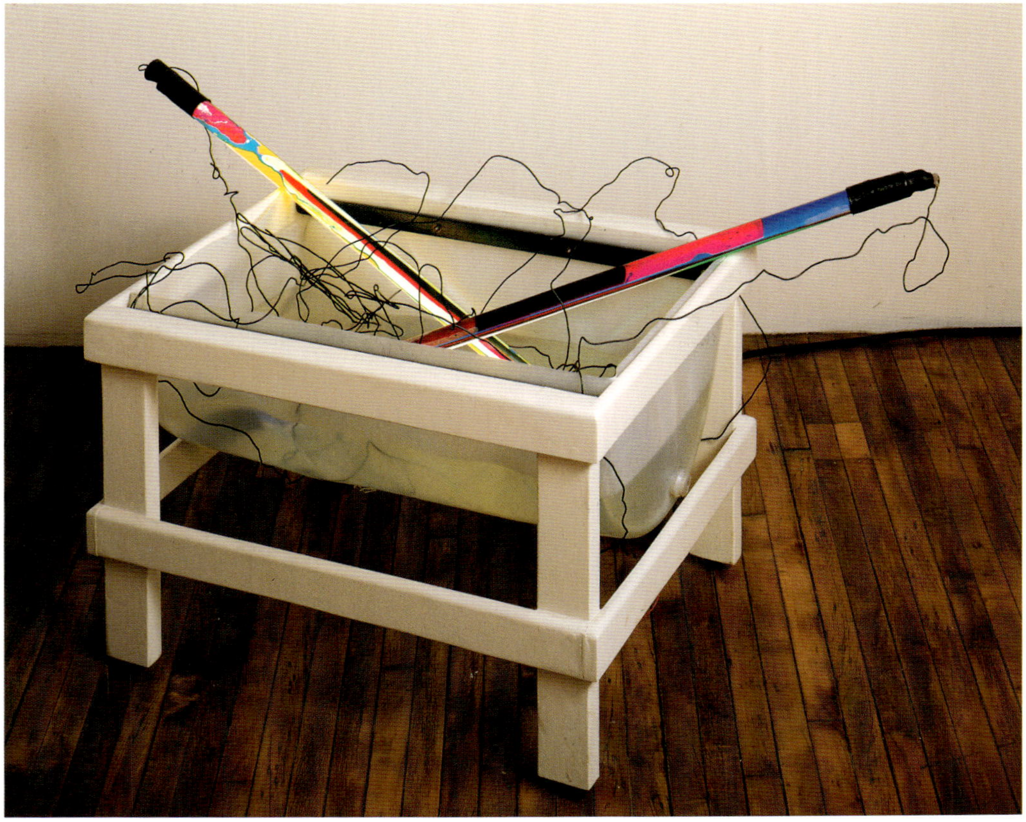

James O. Clark
The Crew
1987

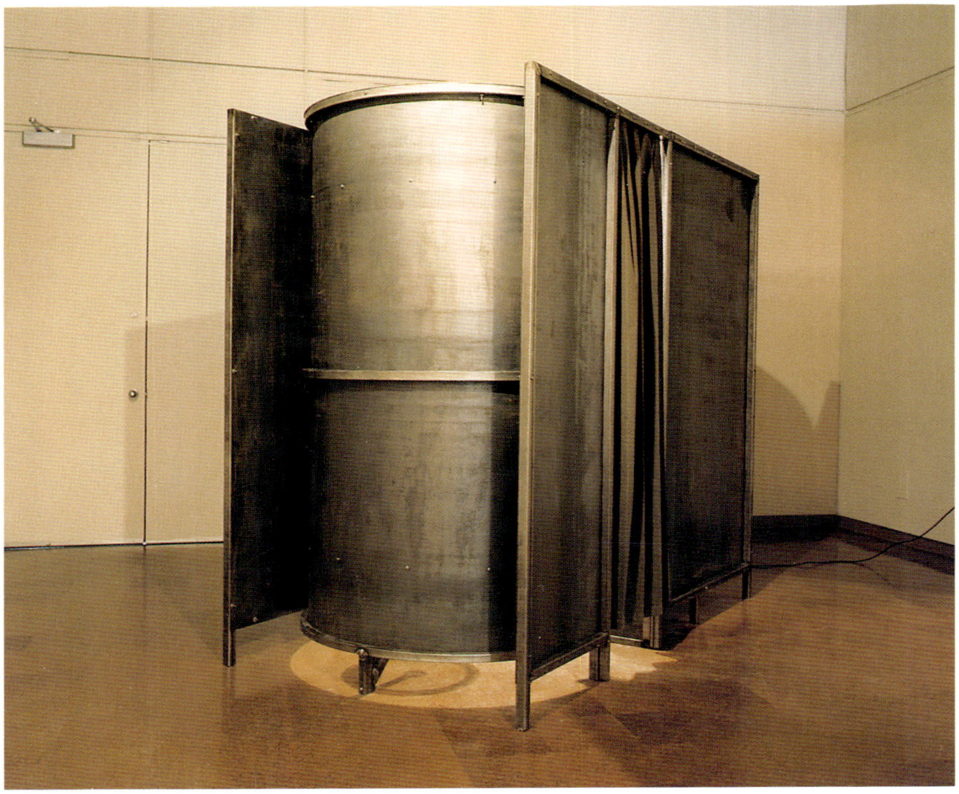

Steve Barry
I.D.
1989

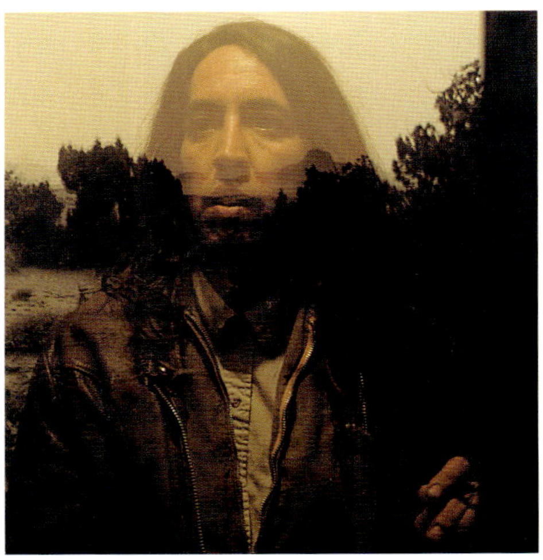

Steve Barry
I.D. (detail showing artist in the process of disappearing)
1989

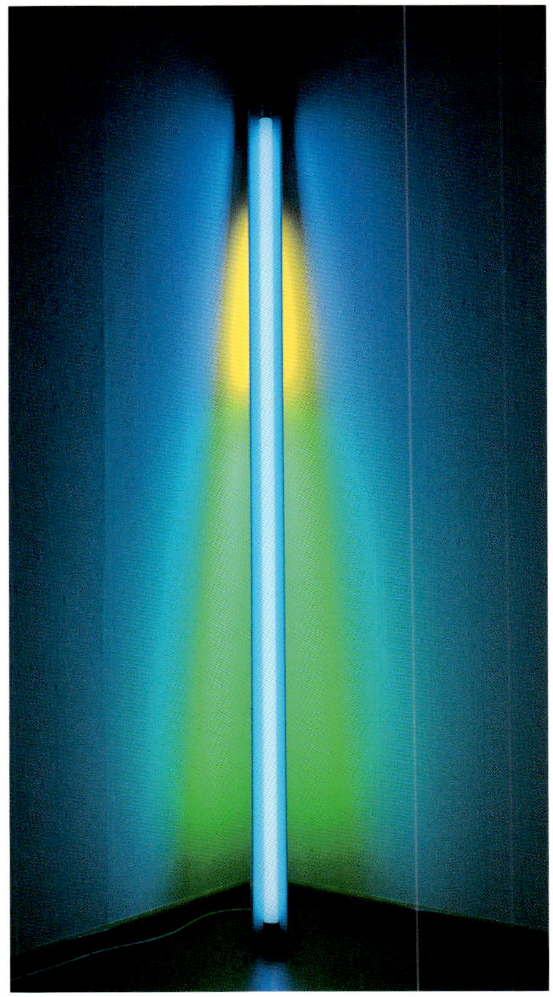

Dan Flavin
Untitled (Fondly, to Helen)
1976

Steven Cortright
Untitled
1988-1990

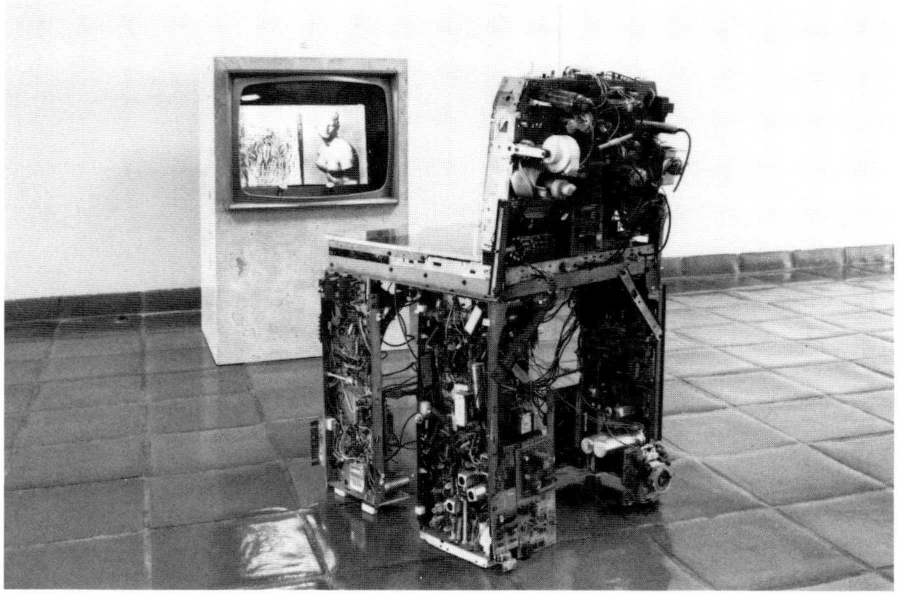

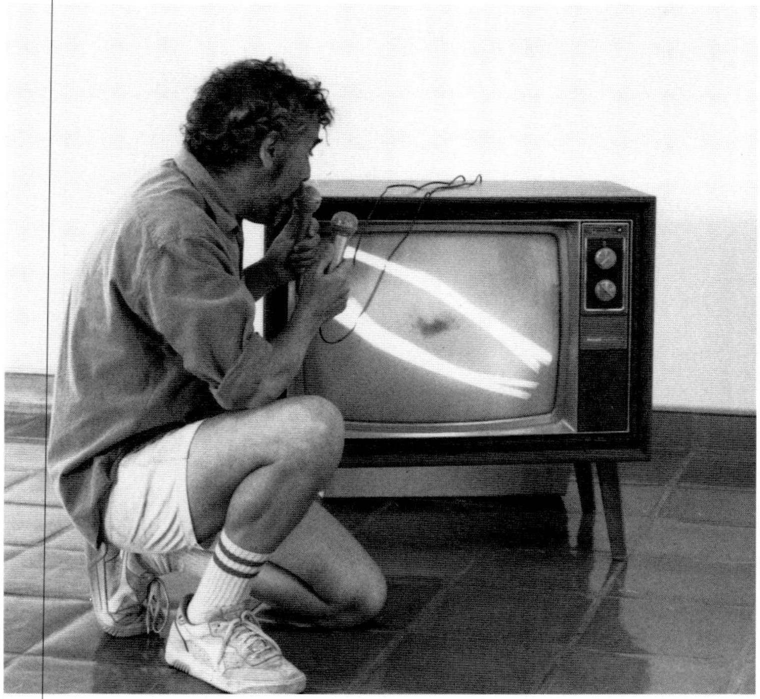

Nam June Paik
Participation TV
1969

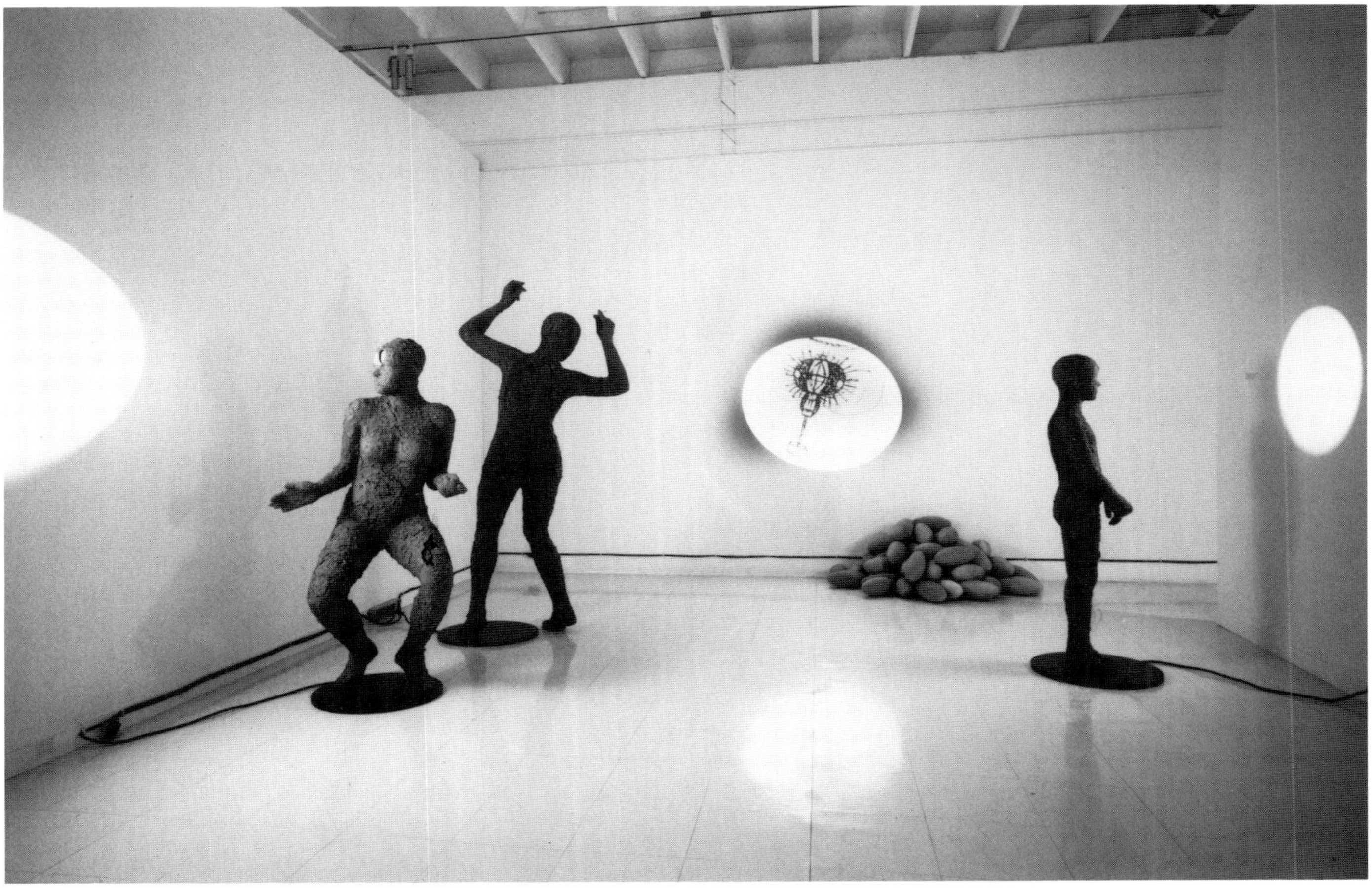

Jonathan Borofsky
Eggs, **1985-1986**
Projector No. 3,214,872, **1989-1990**
Projectors No. 3,214,874 and 75, **1989-1990**

Excerpts from an Interview with Raúl Ruiz

by Jodi Torrent, February 1990

JT: *The Expulsion of the Moors* is your first installation. What brought you to work in this form?

RR: In 1985, when I went back to theater, with my version of Calderon de la Barca's "Life is a Dream" at the Festival of Avignon, I realized that what I did there was very close to a video installation. About the same time, a friend brought me to see a Bill Viola installation and I discovered there the possibility of making a space speak, of giving eloquence to a space—a problem which is at the same time theatrical and cinematographic. . . I am building a bridge between my work in theater and film. . .

JT: What is the subject of *The Expulsion of the Moors*?

RR: . . . the subject of this installation is the total exclusion of one community from another, both living in the same territory. It is not about Christianity nor Islam, but about the complex mechanism that brought apart two communities and of the effort to bring them together. This installation is also about the enormous efforts that, for many centuries, both communities, most of the time without even knowing, have had to make to understand each other. It is more about this than about the justice or injustice of the expulsion, which from the logic of a Catholic system was inevitable. . . this installation allows me to treat an ethical problem in an expressive manner.

JT: Where does Velázquez fit in all this?

RR: Velázquez represents the Spain that closes and cuts herself off from the Spain of the three cultures that I like so much. He is a sort of Master of Ceremony in this process of self-amputation.

JT: How would you define your installation *The Expulsion of the Moors*?

RR: . . . I dislike conceptualization although I like analytical philosophy very much . . . What characterizes an art work is that it is enigmatic and mute, that it is not to be talked about. For many years art was a series of small gestures which caused commentary. This has now been manipulated by the artist as well as the theoreticians searching for things to talk about. As a reaction, some people work in a very closed way, creating works about which one can not say anything. . . my work participates in this type of reaction. I am trying to create a work about which one can not really speak. When I start working on a piece I myself do not understand anything. All are signs, gestures. What they communicate is perplexity. They are not communicating any particular kind of articulate thought or message, though they do transmit a certain ability to do things. Of course that implies much more work than in a traditional contemporary work. I have produced more things, more images, more debris, but all that—at the same time—helps the work to close itself off and does not facilitate a commentary about it.

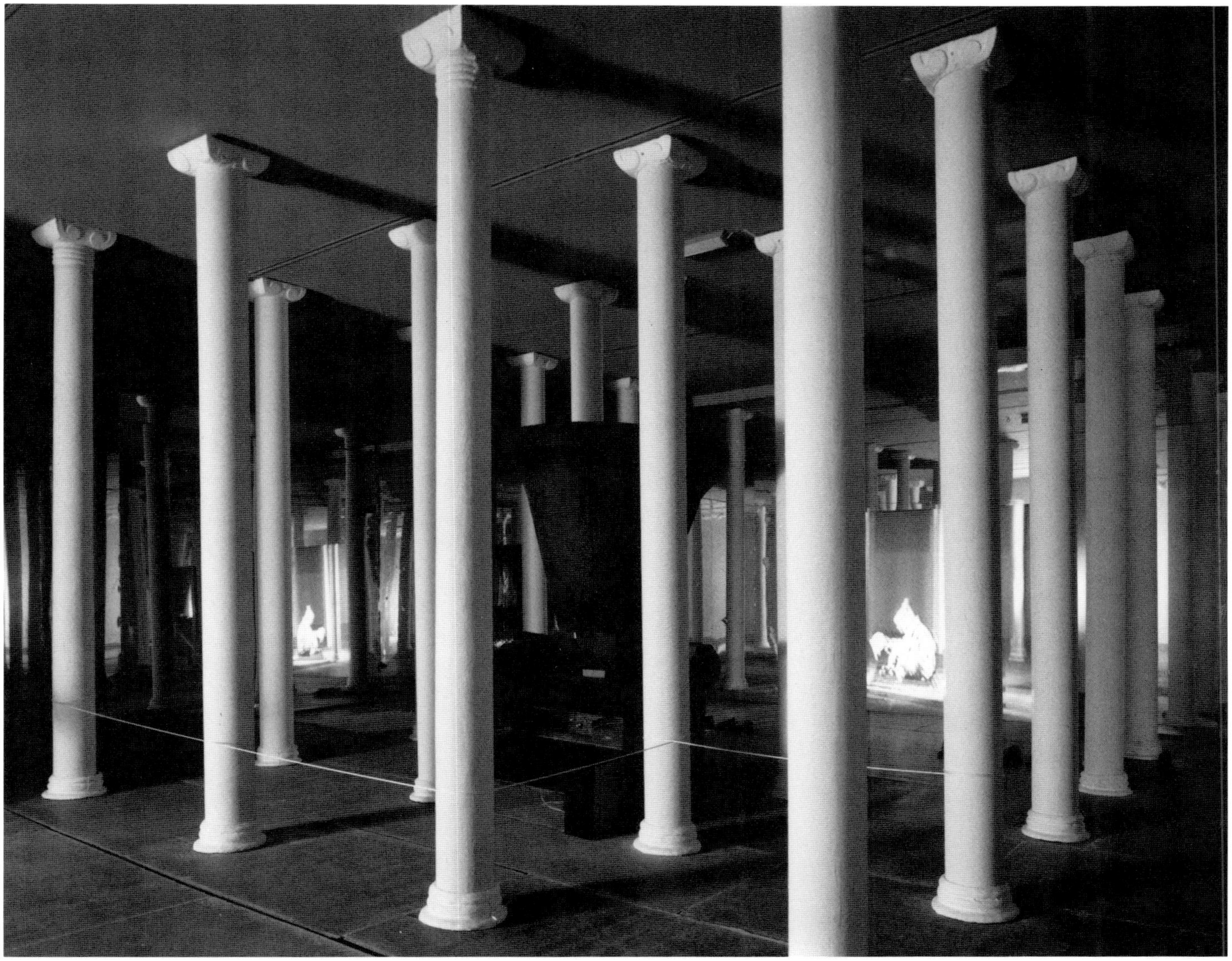

Raúl Ruiz
The Expulsion of the Moors
(installation view at the Institute of Contemporary Art, Boston)
1990

Michael Hardesty
As If
1984

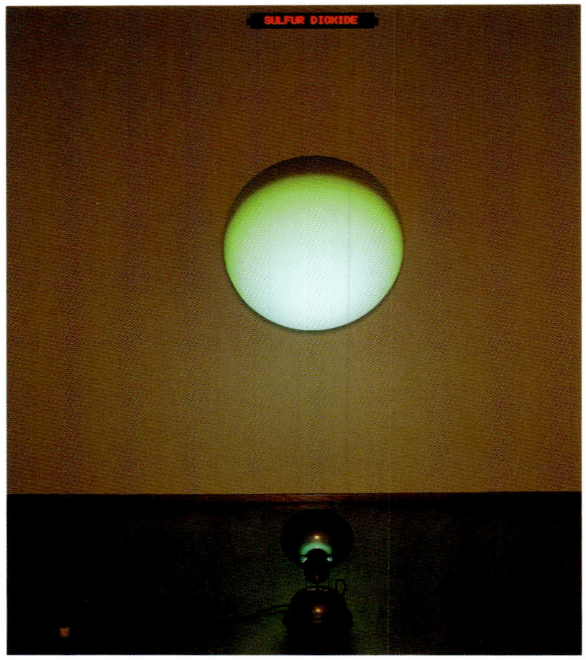

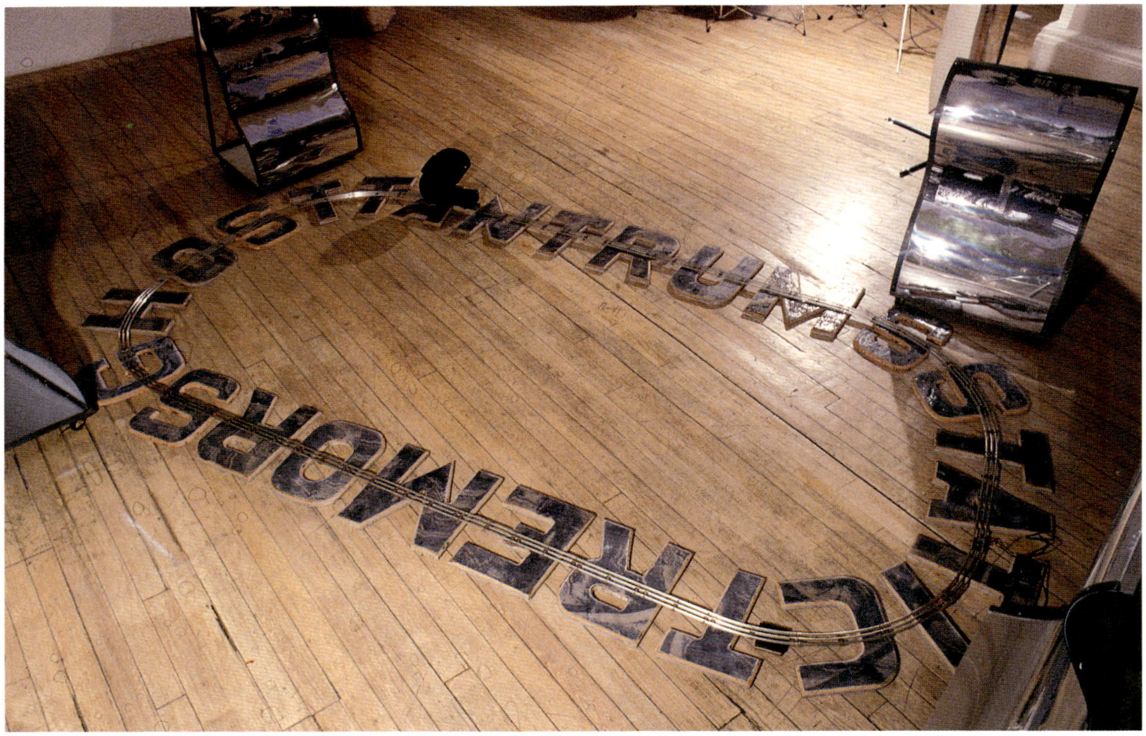

Dennis Oppenheim
Ghost Tantrums, Static Tremors
1988-89

Roland Brener
Small Talk
1987

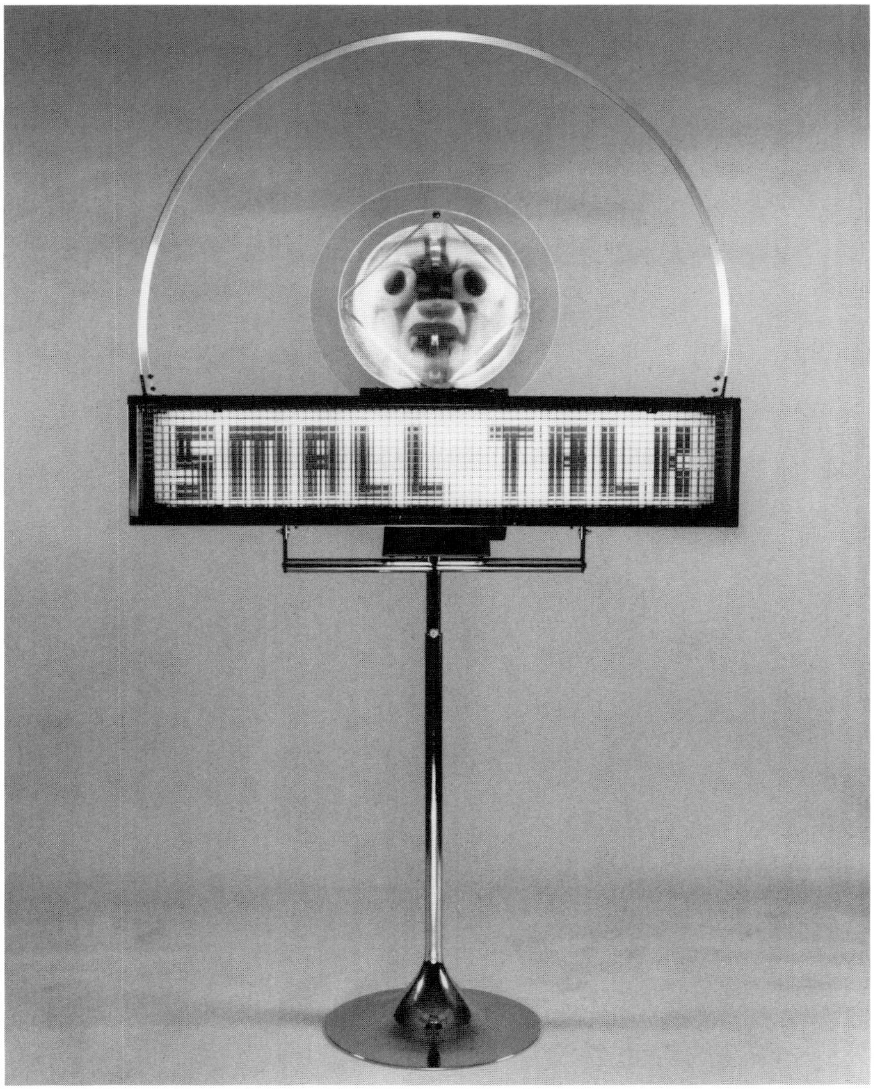

Joe Jones
Music Machine
1976

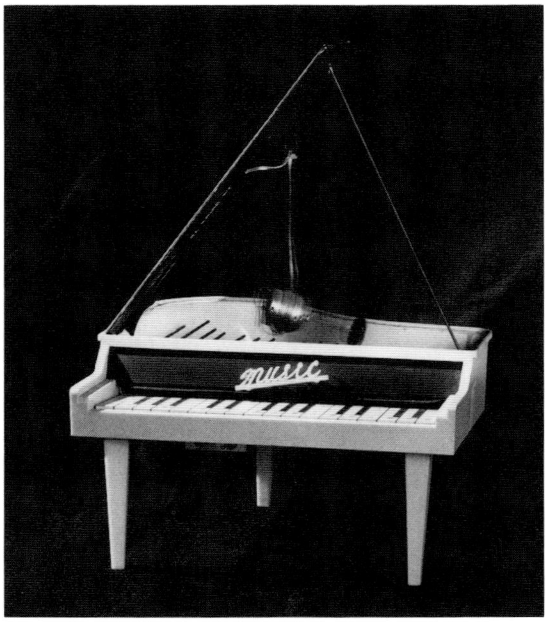

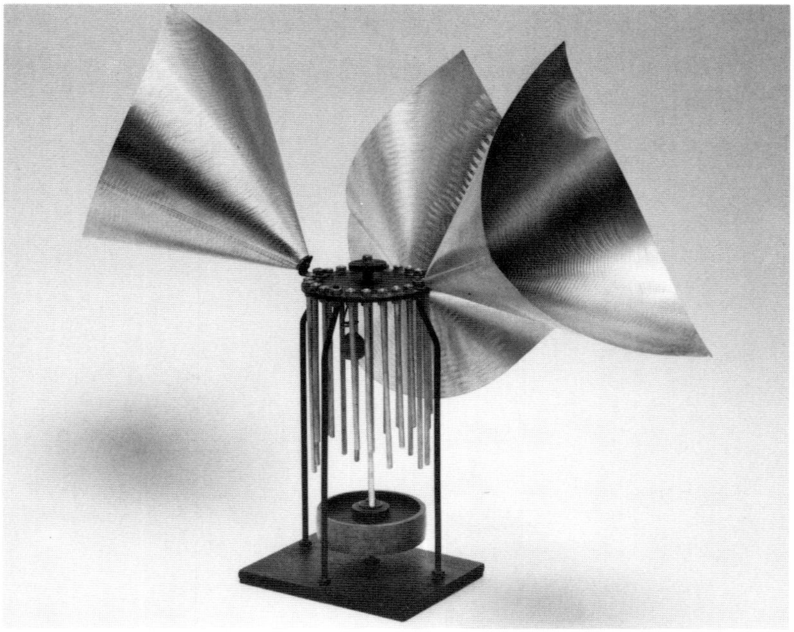

François and Bernard Baschet
Musical Sculpture No. 20
1967

Takis
Musicale
1965-1968

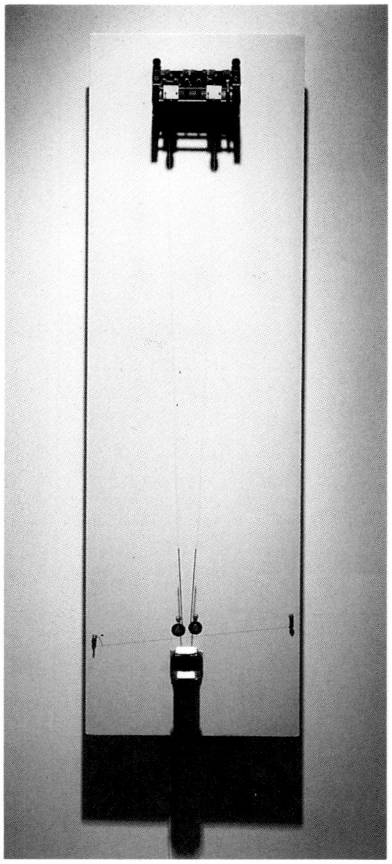

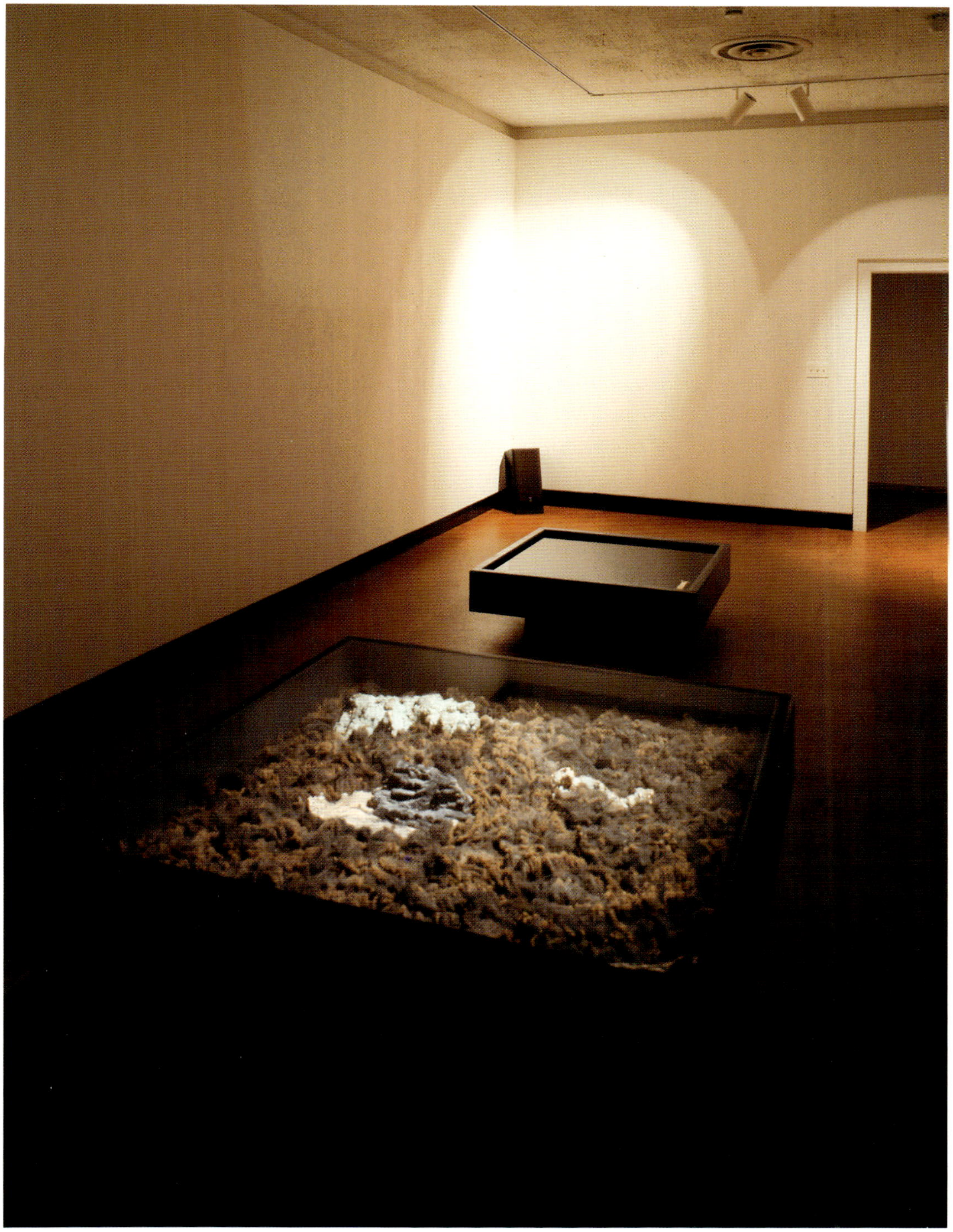

Liz Phillips
Soundtable I and II
(Wet and Dry Landscape)
1989-1990

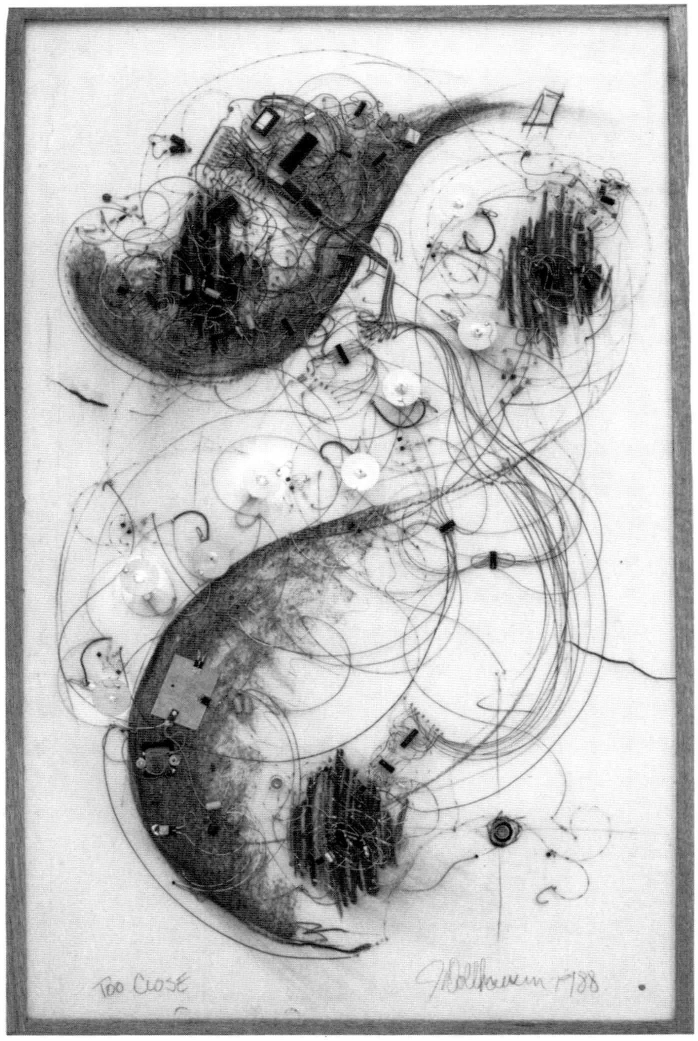

Jack Dollhausen
Too Close
1988

Bill Bell
Where Are We?
1988

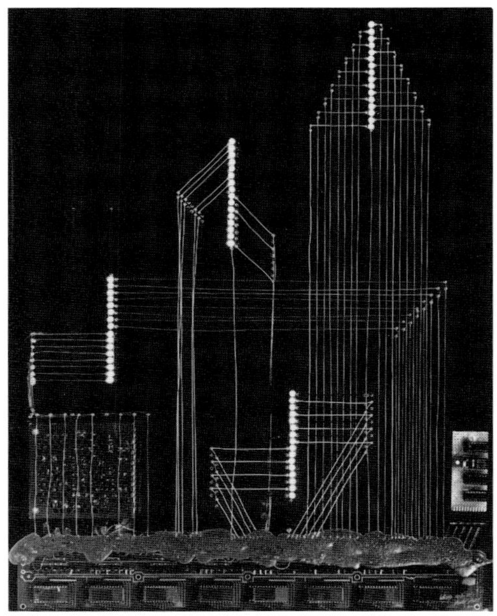

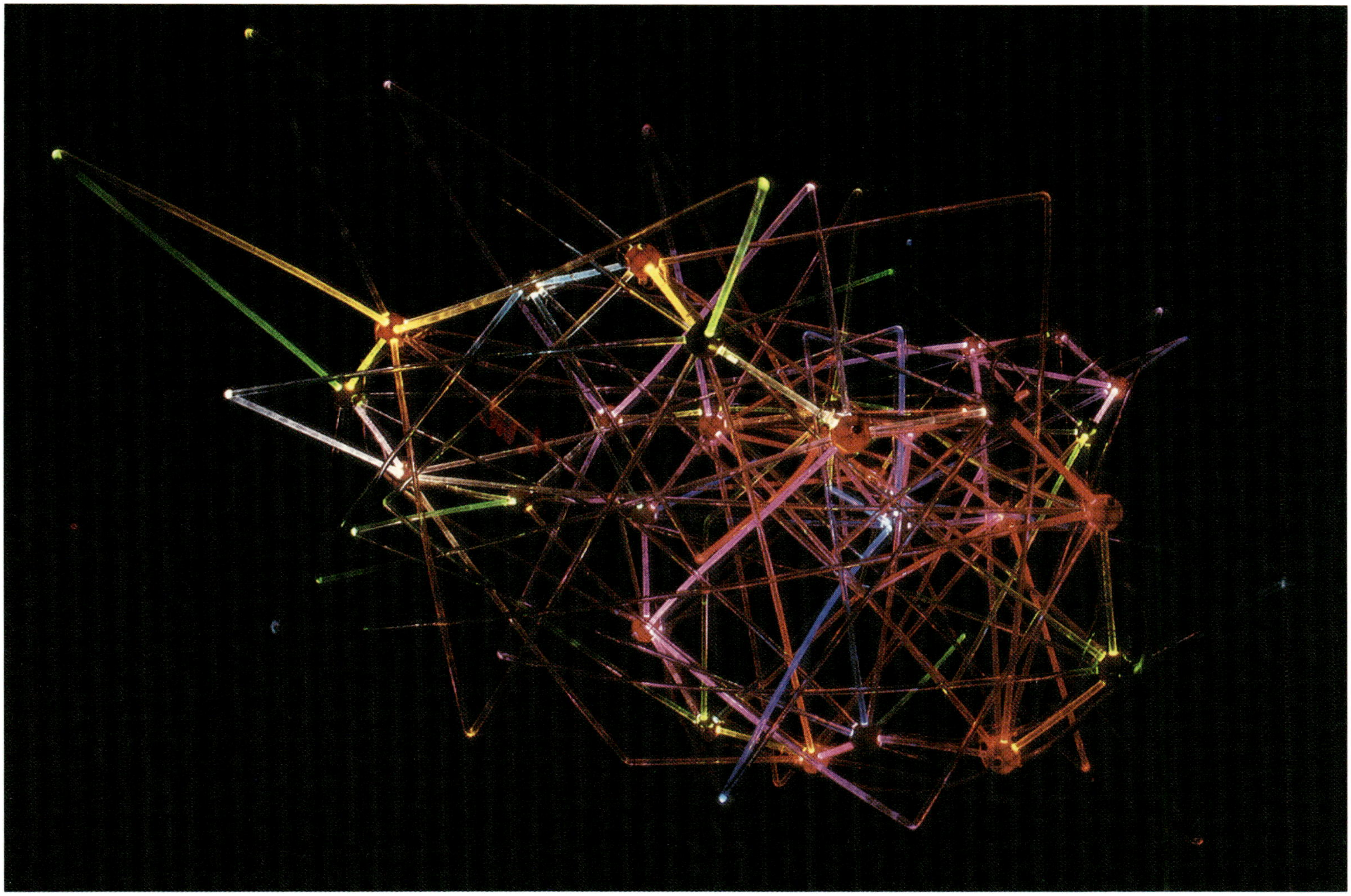

Milton Komisar
The Philosopher's Geometry
1988

Peter Vogel
Two Different Series
1979

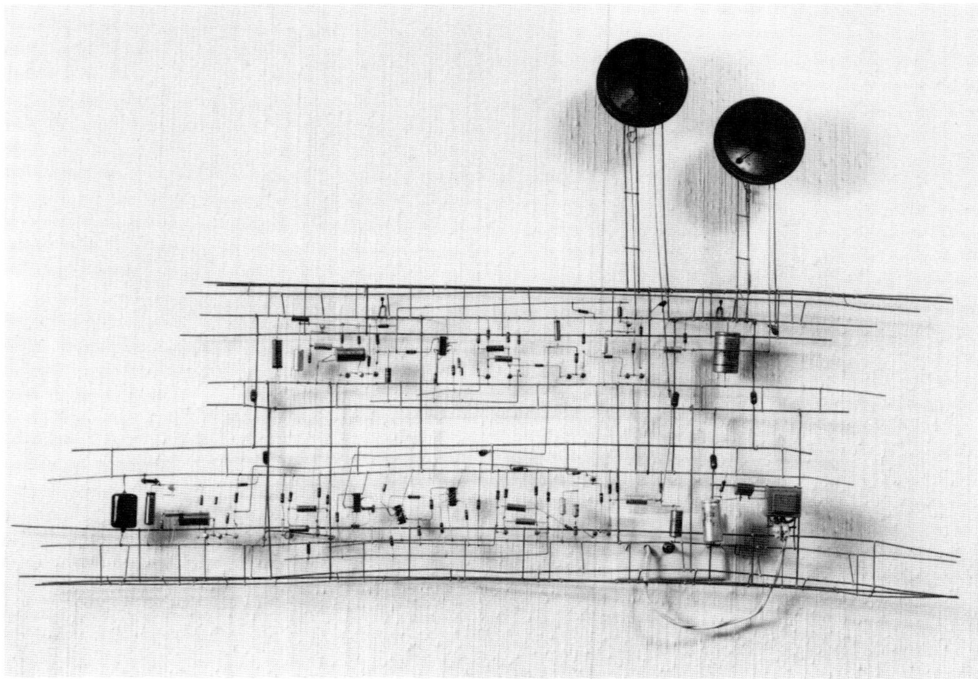

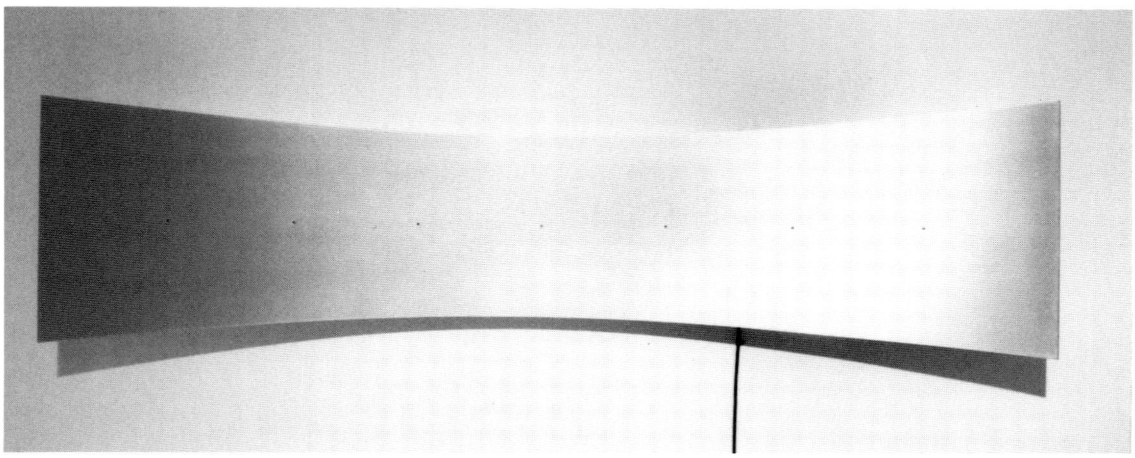

Howard Jones
Sonic Six
1967-1968

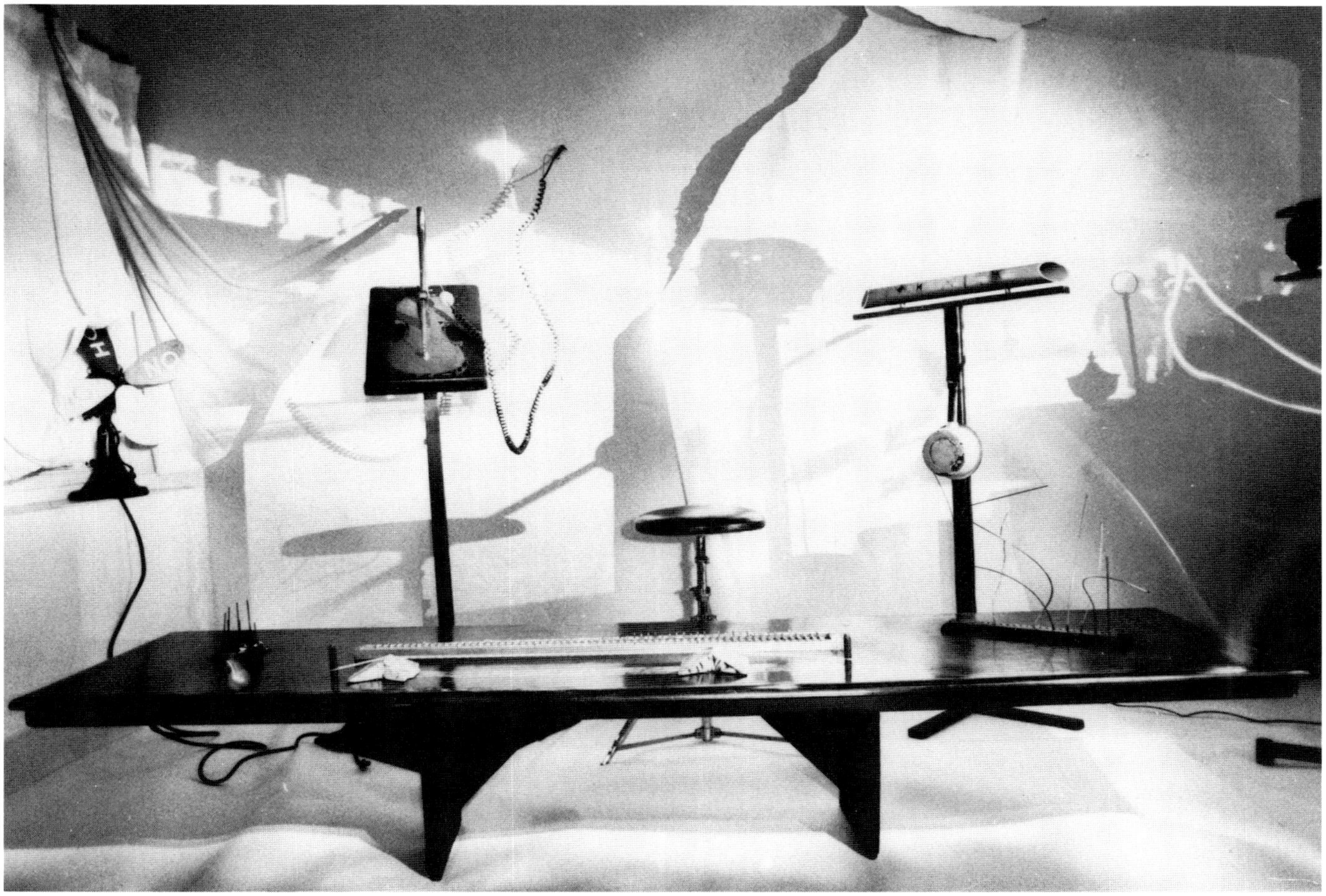

Richard Dunlap
Idiophonic Landscape
1990

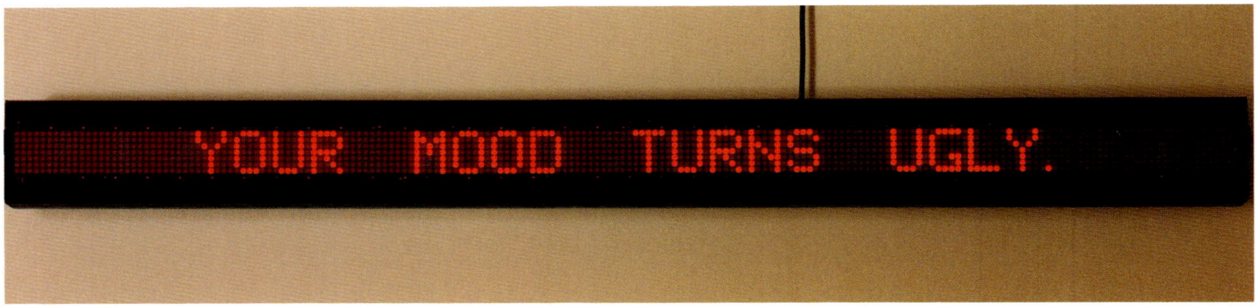

Jenny Holzer
More Survival
1985

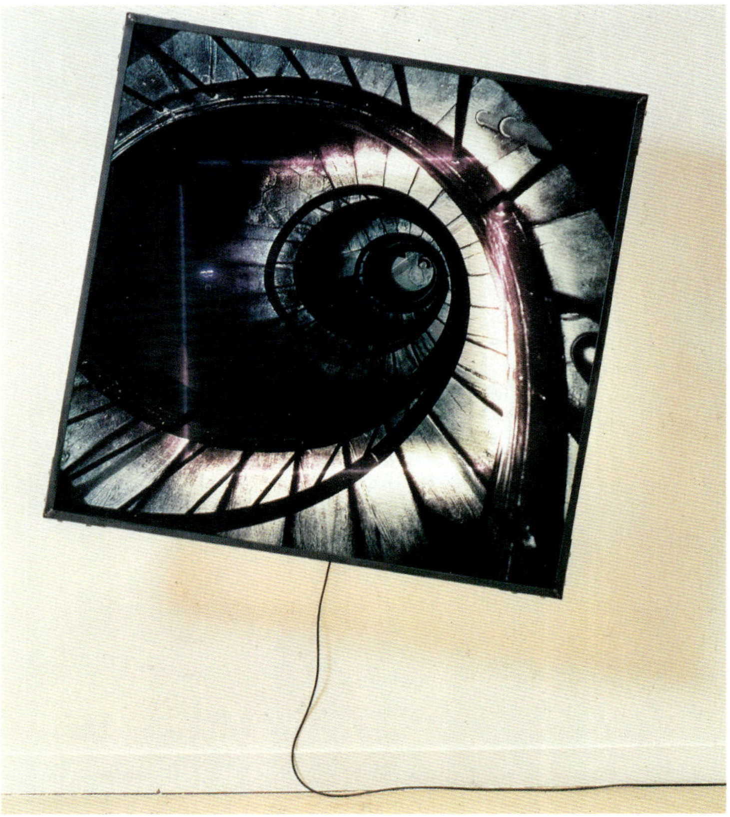

Jon Kessler
16 Rue Chanoinesse
1989

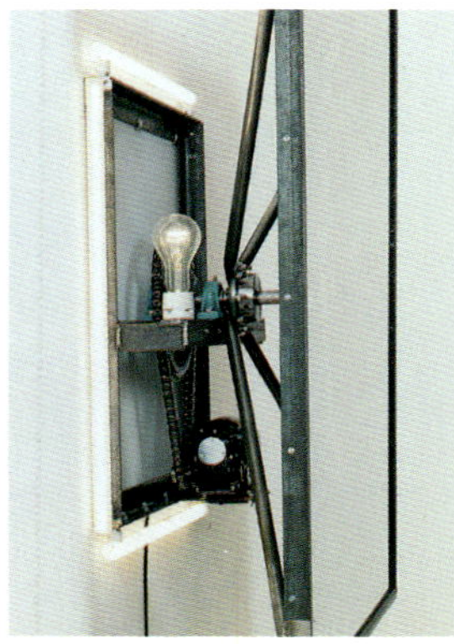

Jon Kessler
16 Rue Chanoinesse (side view)
1989

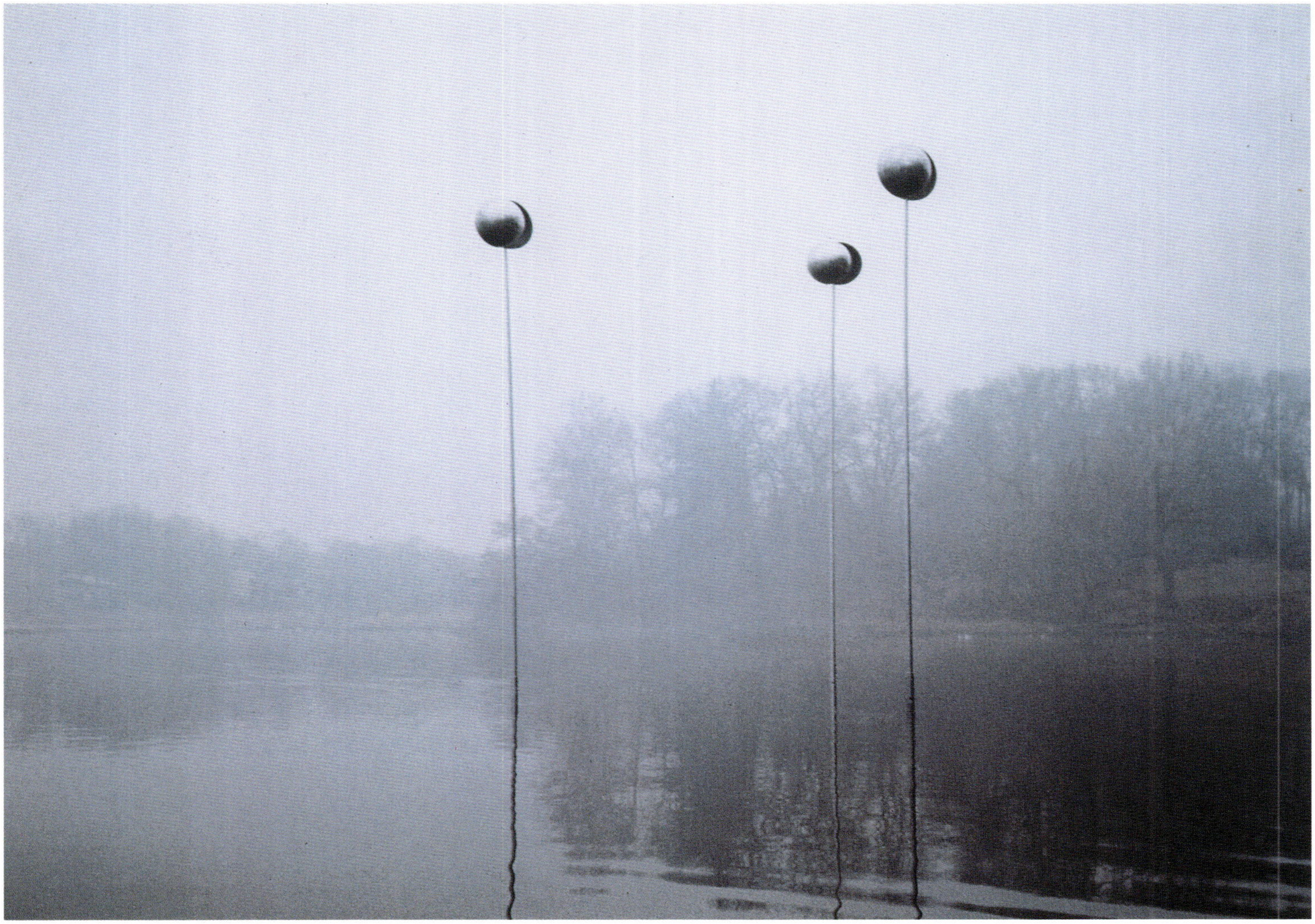

Harold Lehr
Water Sculptures
1972

Dan Flavin
Monument for V. Tatlin
1969

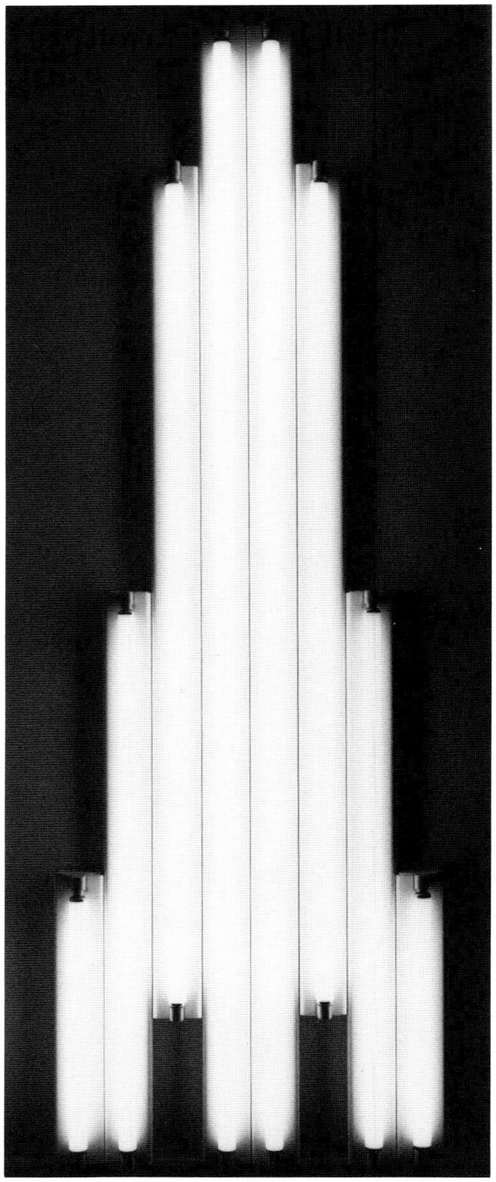

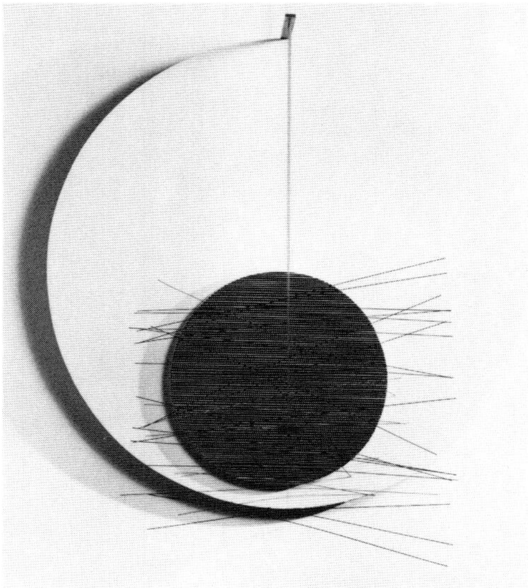

Jesus-Raphael Soto
Rond Blanc
1970

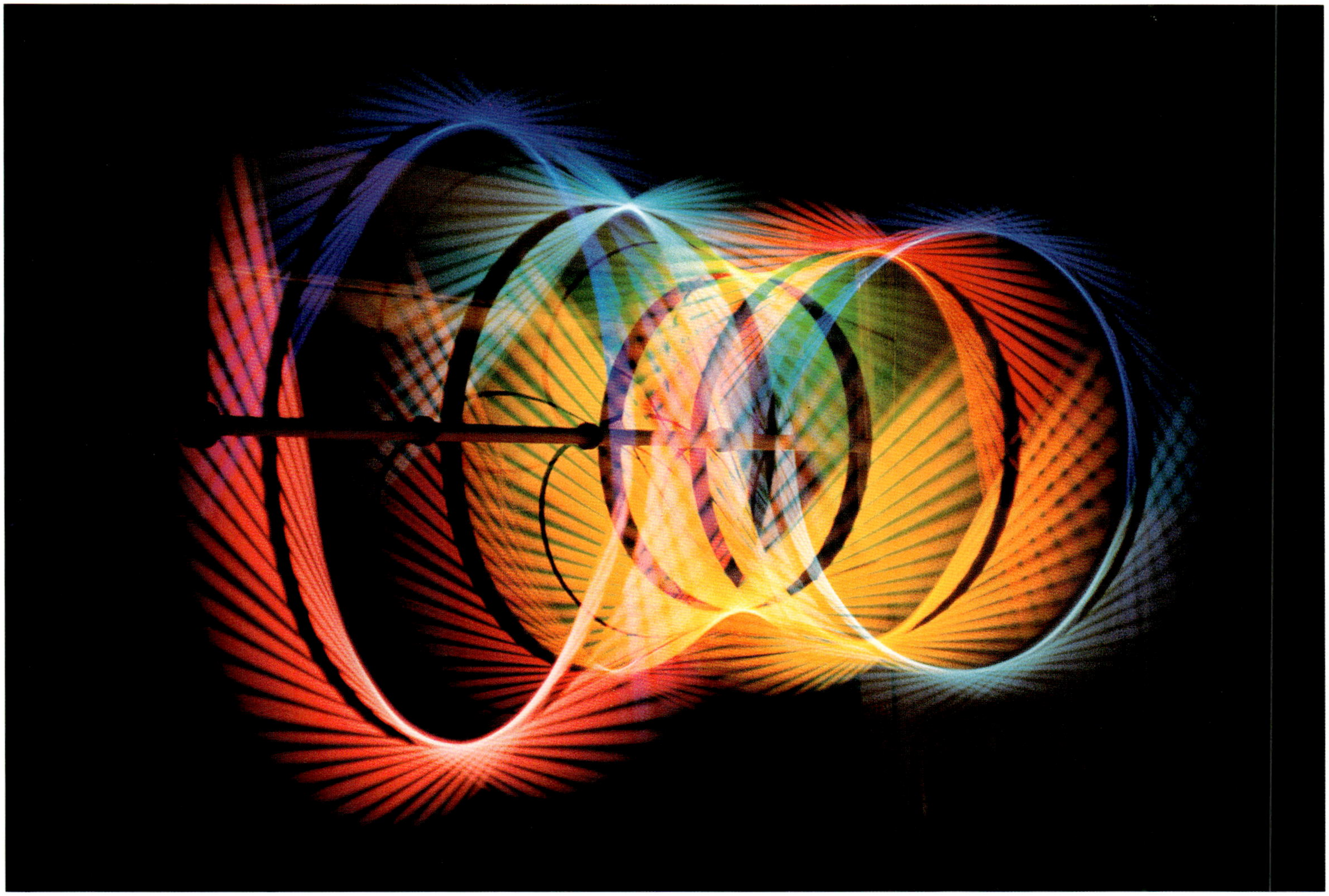

Alejandro and Moira Siña
Spinning Shaft
1984

Alan Rath
Voyeur II
1989

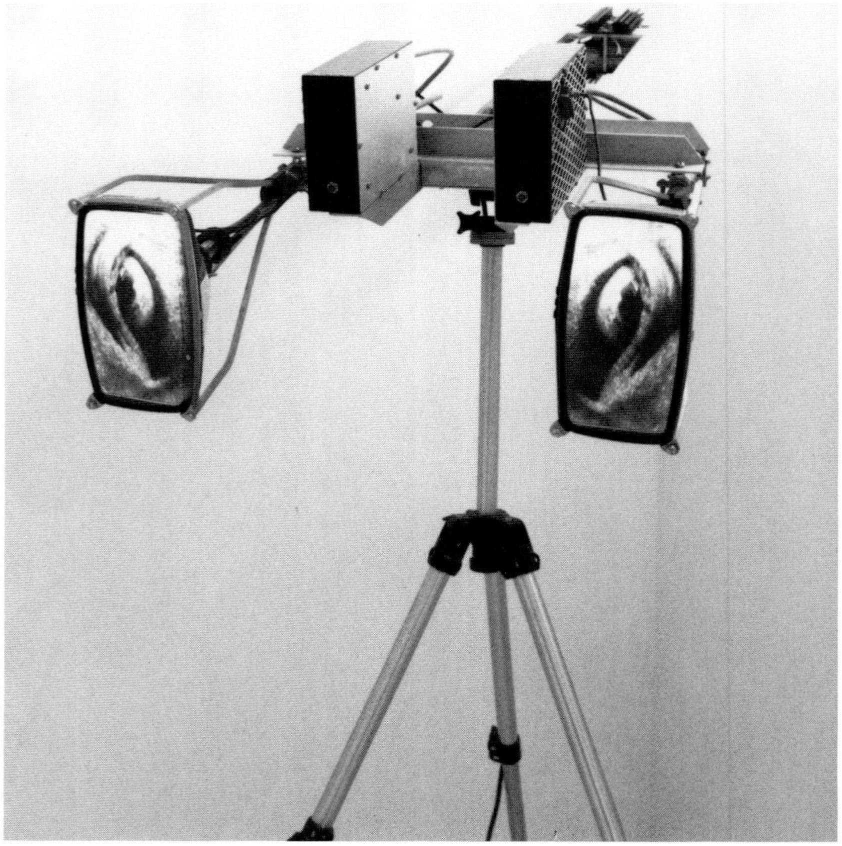

Jack Dollhausen
Midwest Summer Night's Dream
1977

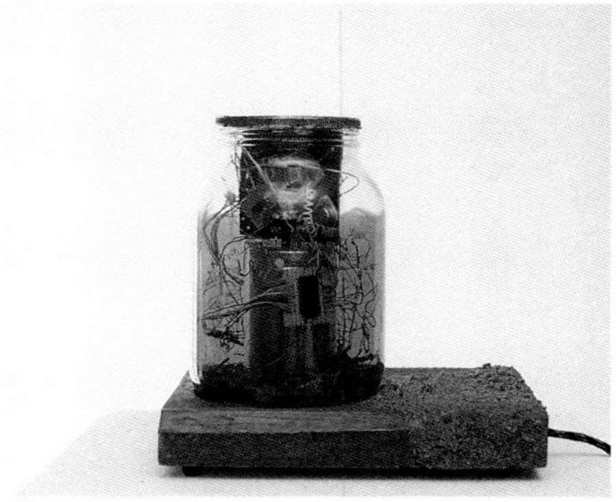

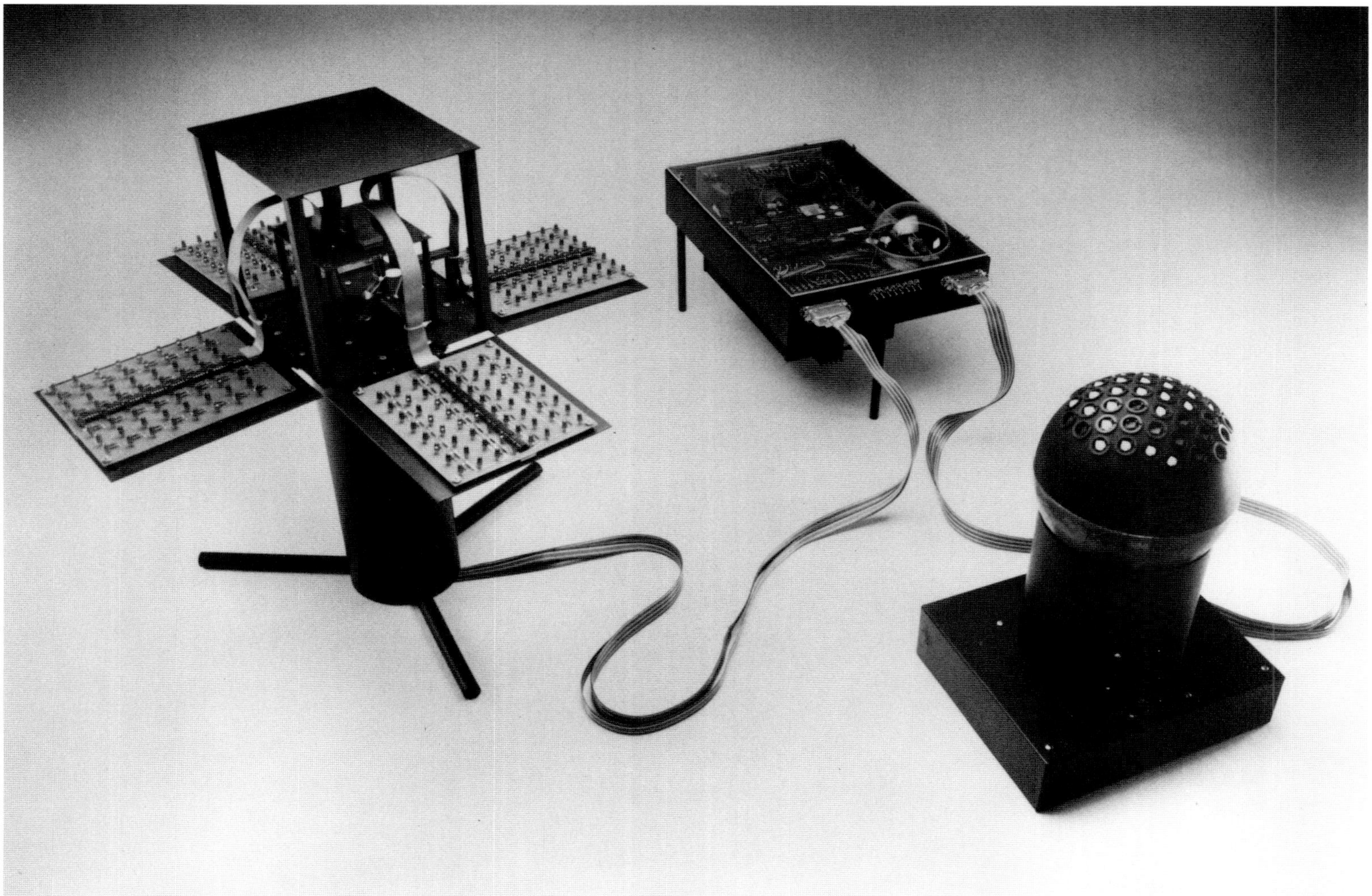

James Seawright
House Plants
1984

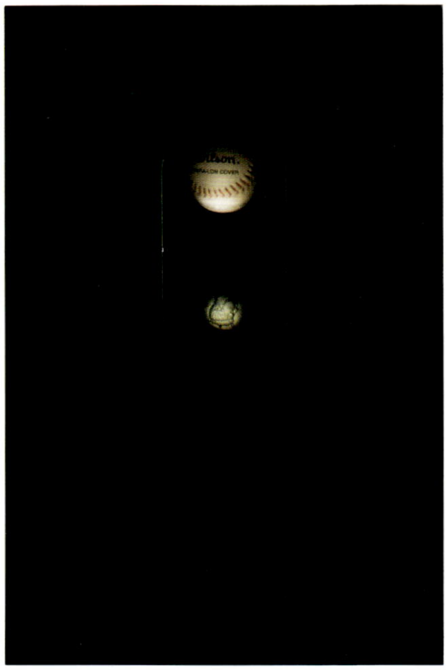

Ted Victoria
Hey Pete, Heads or Tails
1989

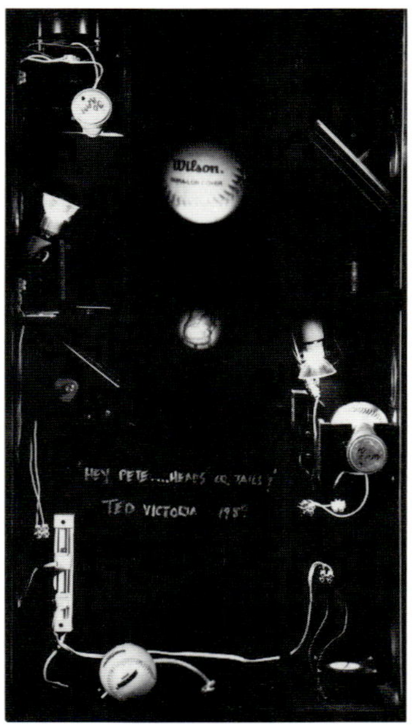

Ted Victoria
Hey Pete, Heads or Tails
(detail of interior)
1989

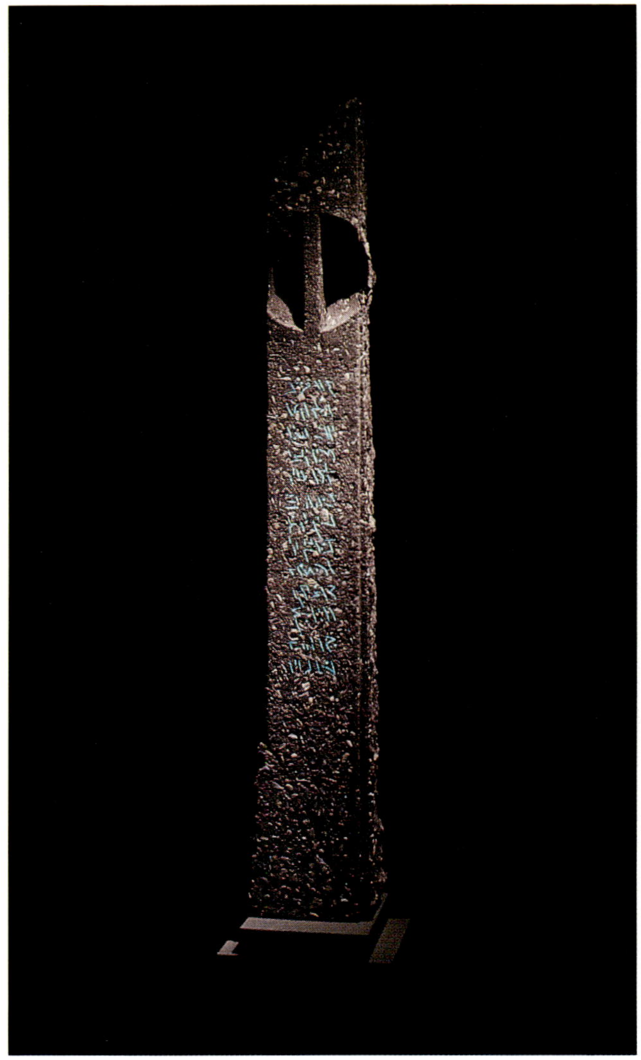

Clyde Lynds
Stele LXXV, Gnomon
1989

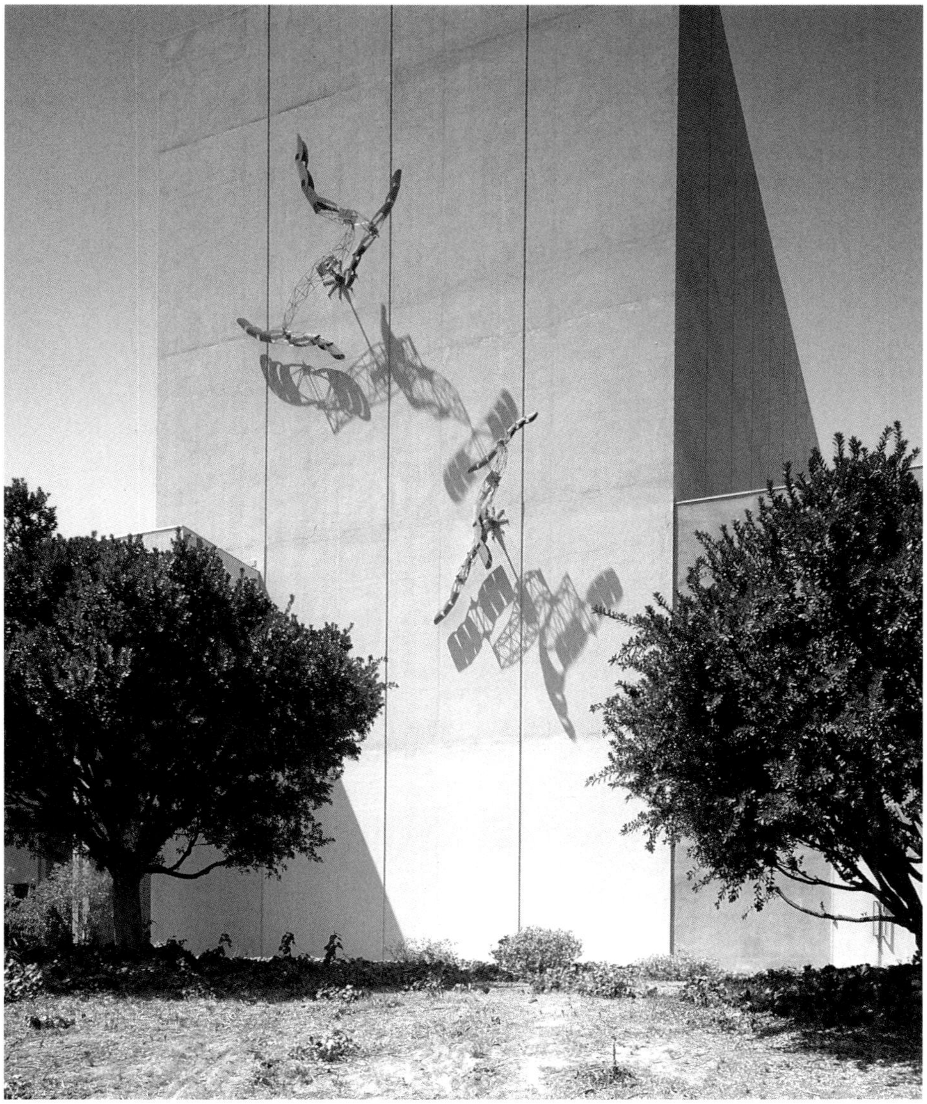

Evan Lewis
Quiet Storm
1990

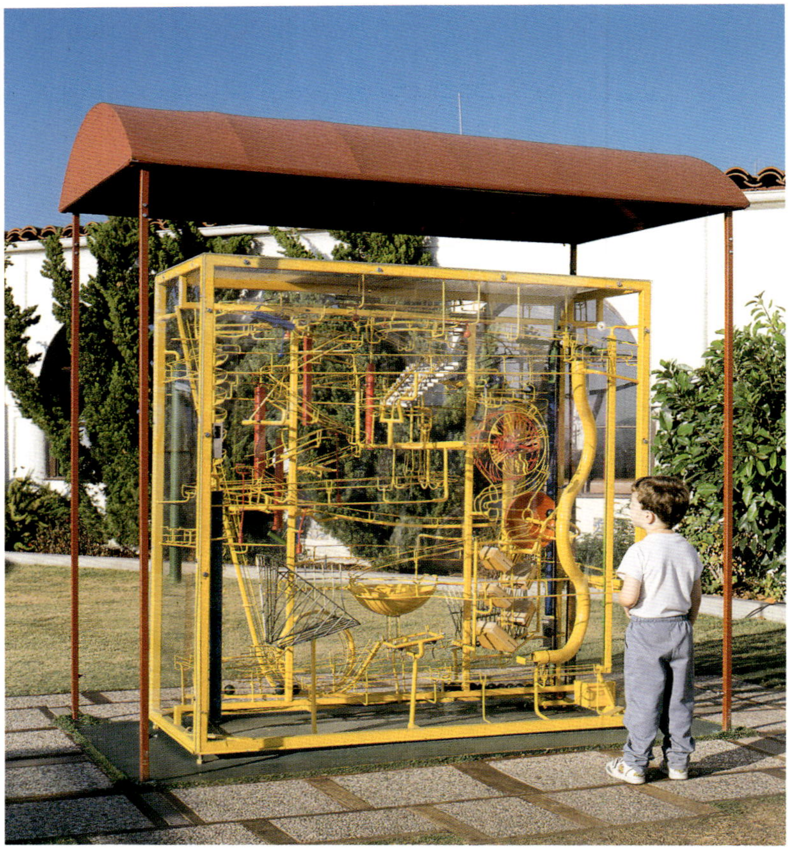

George Rhoads
Goodtime Clock IV
1984

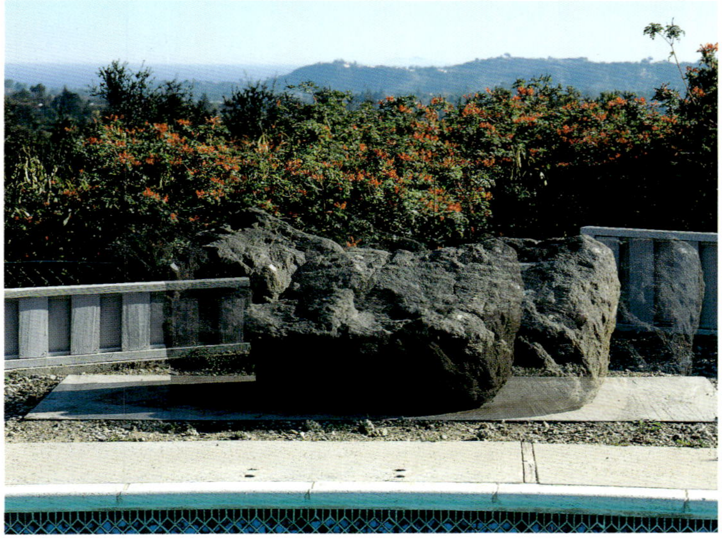

Rikuro Okamoto
Moving Rock No. 22
1988

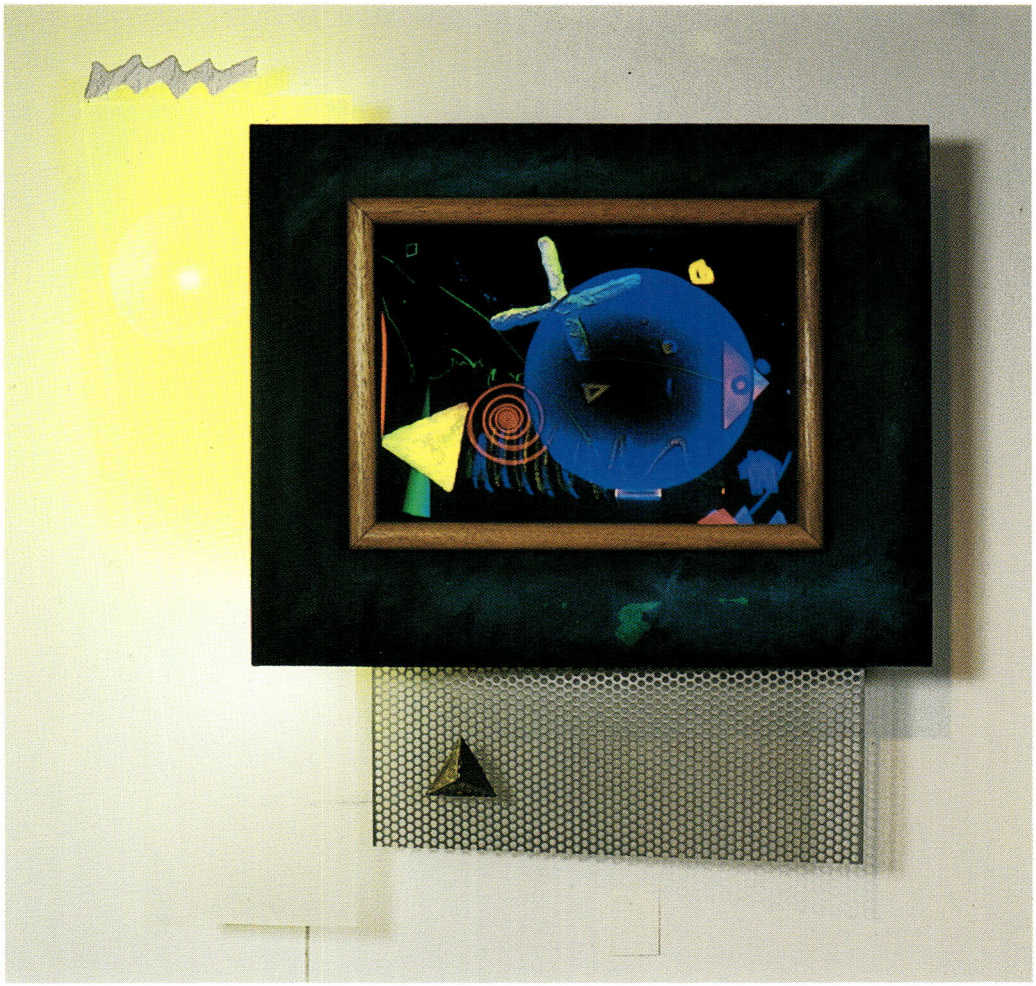

Art Spellings
Free Spirit
1987

George Rickey
Four Trapezoids as Two Rectangles II
1987

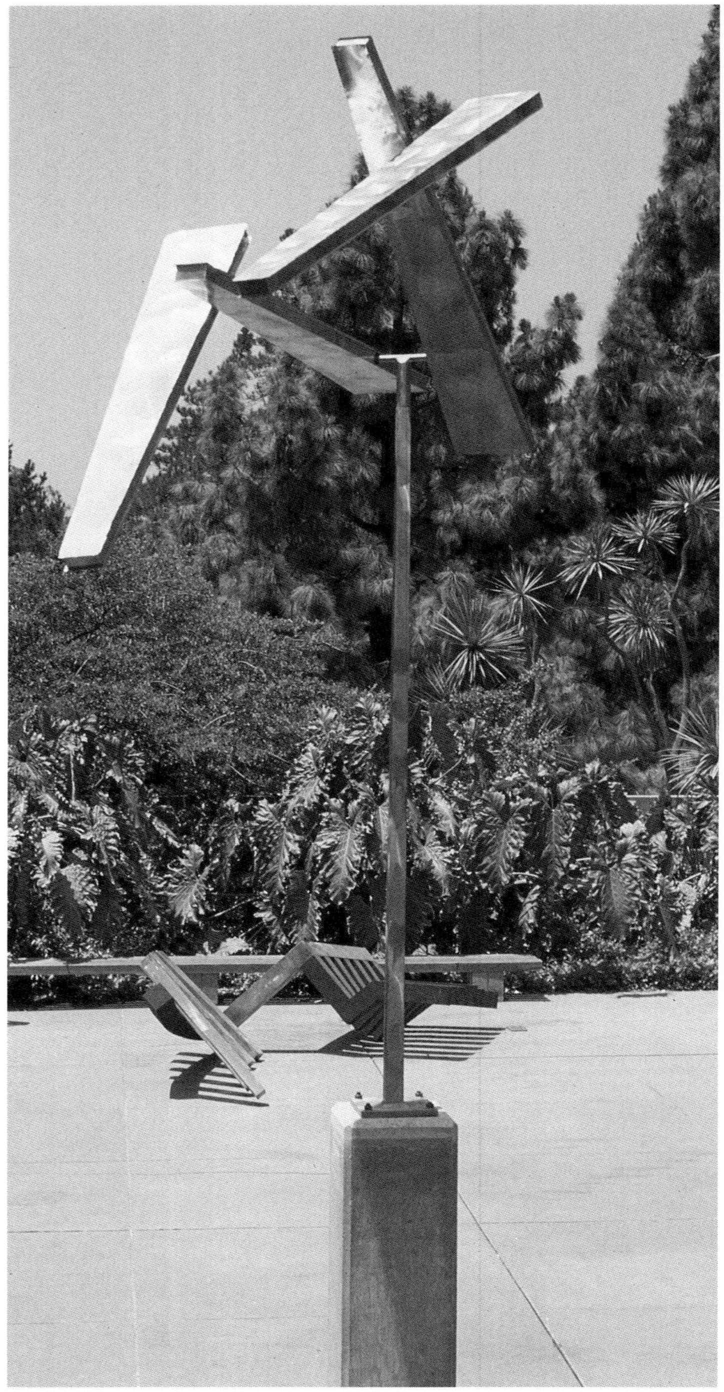

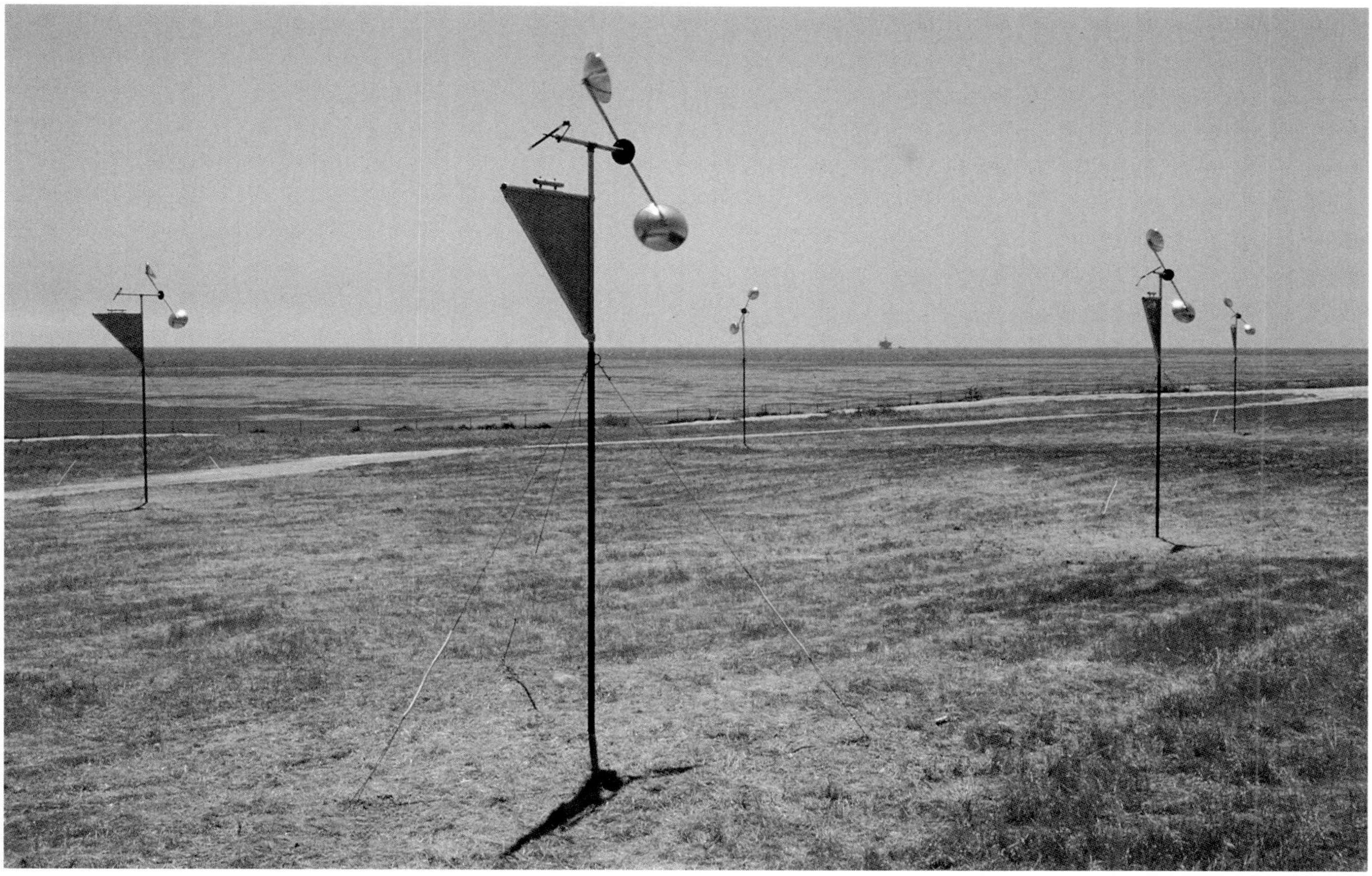

Bill and Mary Buchen
Wind Gamelan
1987

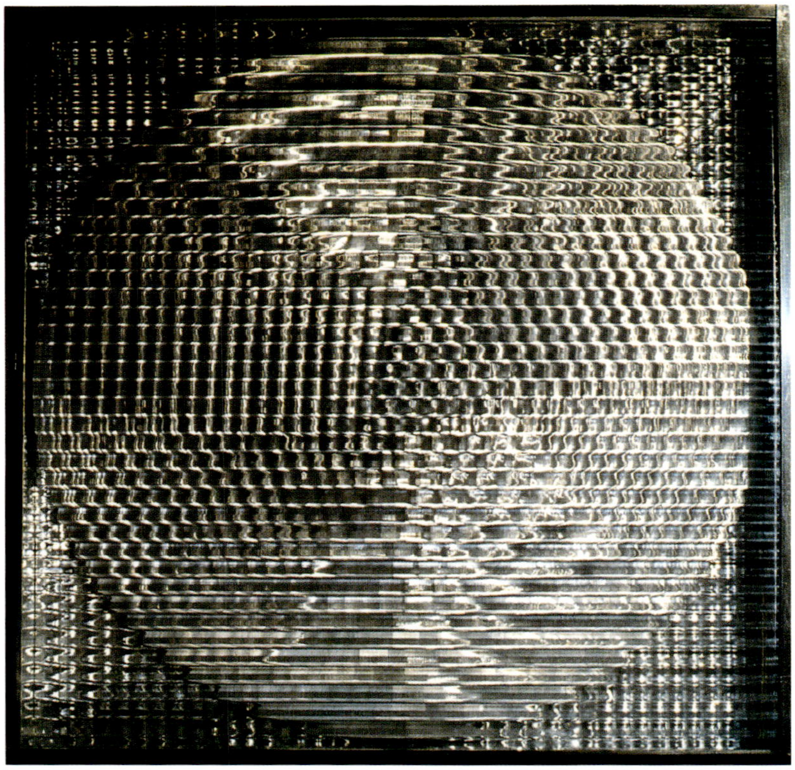

Heinz Mack
Light Dynamo No.1
1966

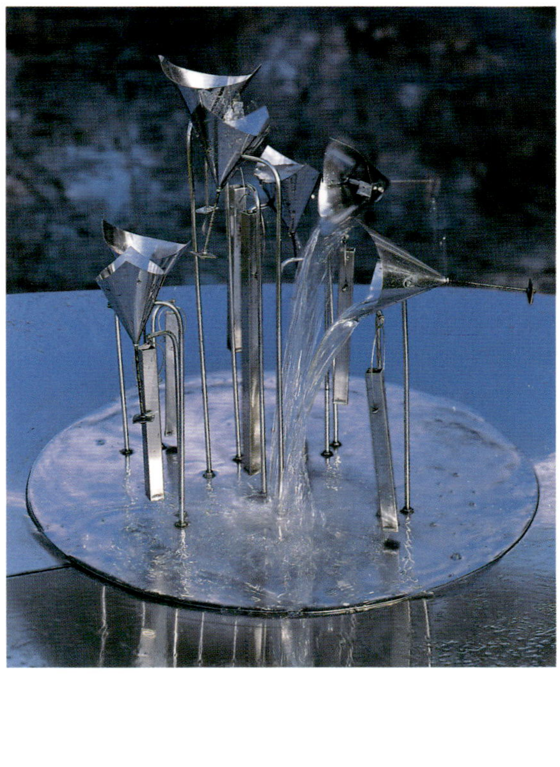

François and Bernard Baschet
Fountain
1974

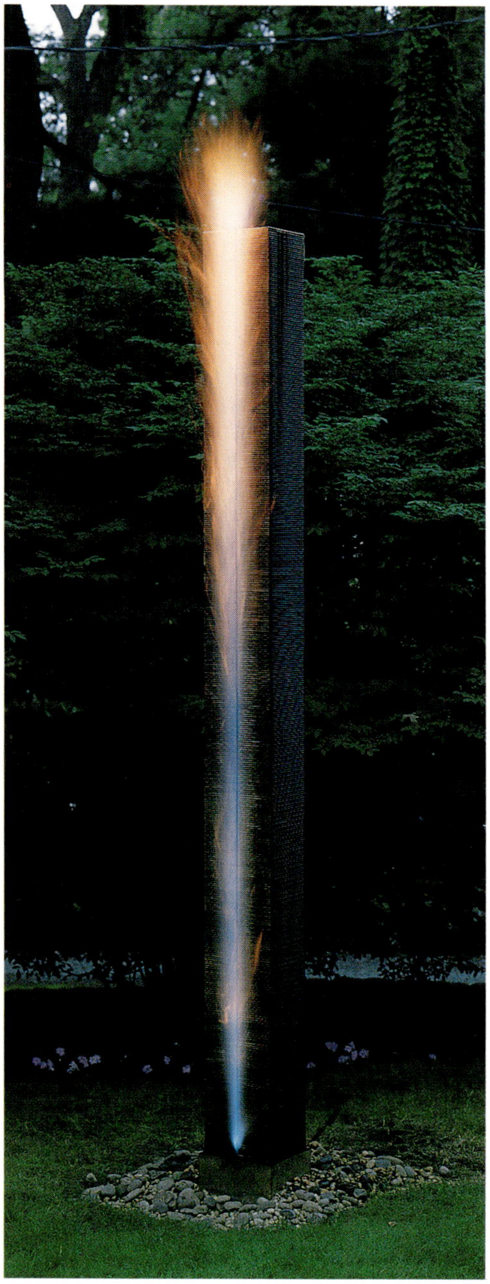

Eric Orr
Prime Matter VIII
1987

Checklist of Works

Unless otherwise indicated, all dimensions are in inches; height precedes width precedes depth.

Alice AYCOCK

Born 1946, Harrisburg, Pennsylvania
B.A., Douglass College, New Brunswick, New Jersey, 1968
M.A., Hunter College, 1971
Lives in New York City

1 *Fata Morgana (Of Things Seen in the Sky)* 1984
mixed media 149 x 57 x 130
Courtesy John Weber Gallery, New York

Steve BARRY

Born 1956, Jersey City
B.F.A., School of Visual Arts, New York, 1980
M.F.A., Hunter College, New York, 1984
Lives in Corrales, New Mexico

2 *I.D.* 1989
steel, mirror, plastics, light, canvas, photograph, electronics
91 x 48 x 96
Lent by the artist

François and Bernard BASCHET

Born 1920 and 1917, Paris, France
Live in Paris

3 *Musical Sculpture No. 20* 1967
brushed stainless steel, aluminum 22 1/4 x 34
Collection University Art Museum, Santa Barbara

4 *Fountain* 1974
stainless steel and electric pump 33 x 33 x 54
Collection David W. Bermant

Bill BELL

Born 1928, Pittsburgh
B.A., Princeton University, 1949
Lives in Brookline, Massachusetts

5 *Where Are We?* 1988
plexiglas, computer, LEDs 24 1/2 x 2 x 19
Collection National Shopping Centers, Inc., Harrison, New York

Harry BERTOIA

Born 1915, San Lorenzo, Italy;
died 1978, Bally, Pennsylvania
Cranbrook Academy of Art, 1937-1939

6 *Sound Sculpture* 1973
beryllium, copper, bronze 96 x 8 x 8
Collection National Shopping Centers, Inc., Harrison, New York

Jonathan BOROFSKY

Born 1942, Boston
B.F.A., Carnegie Mellon University, 1964
M.F.A., Yale University, New Haven, 1966
Lives in Venice, California

7 *Eggs* 1985-1986
cast urethane 54, each 6 x 12
Collection Robert and Mary Looker

8 *Projector No. 3,214,872* 1989-1990
fiberglass, celluclay, resin, projector, steel 55 3/8 x 20 diameter
Collection Robert and Mary Looker

9 *Projector No. 3,214,874* 1989-1990
fiberglass, celluclay, resin, projector, steel, wood
73 x 38 diameter and 58 x 29 1/2 diameter
Courtesy Paula Cooper Gallery, New York

10 *Projector No. 3,214,875* 1989-1990
fiberglass, celluclay, resin, projector, steel, wood
58 x 29 1/2 diameter
Courtesy Paula Cooper Gallery, New York

Martha BOTO

Born 1925, Buenos Aires
Ernesto de la Cracova Superior Academy of Fine Arts, Buenos Aires, 1944-1946
Lives in Paris

11 *Variations Lumineuses, Paris* 1965
wood, plexiglas, mirror, incandescent lights 32 x 27 x 23
Collection David W. Bermant

Roland BRENER

Born 1942, Johannesburg, South Africa
St. Martin's School of Art, London, 1963-1966
Lives in Victoria, British Columbia

12 *Small Talk* 1987
mixed media 84 x 42 x 12
Collection Art Gallery of Greater Victoria, Vancouver
Volunteer Committee Purchase Fund

13 *Life in Quick Basic* 1987-1988
steel, plastic, electro-mechanical hardware and computer
76 x 59 1/2 x 58 1/2
Collection Vancouver Art Gallery

Bill and Mary BUCHEN

Born 1949, Minneapolis, and 1948, Jamestown, North Dakota
Live in New York City

14 *Singer Songs* 1980
mixed media, sound installation 53 x 46 1/2 x 21
Lent by the artists

15 *Wind Gamelan* 1987
mixed media, sound installation
Lent by the artists

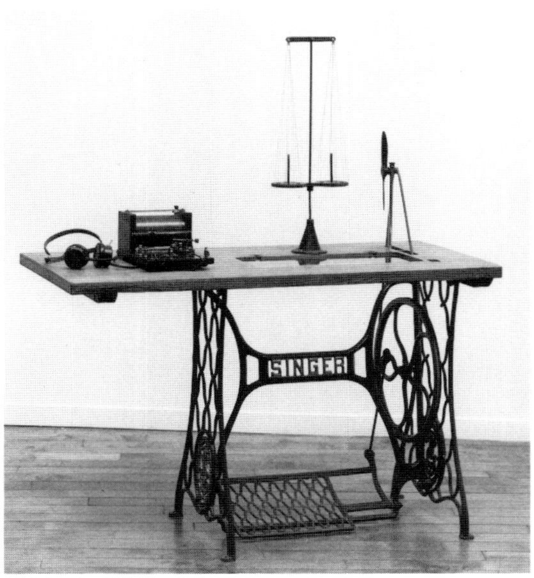

Bill and Mary Buchen
Singer Songs
1980

Alexander CALDER

Born 1898, Philadelphia; died 1976, New York City
M.E., Stevens Institute of Technology, Hoboken, 1918

16 *Blue Wing, Black Wing* 1965
paint on metal 24 x 30 x 32
Private Collection

James O. CLARK

Born 1948, Coatesville, Pennsylvania
B.S., Kutztown University of Pennsylvania, 1974
Lives in Brooklyn

17 *The Crew* 1987
mixed media 24 x 36 x 42
Lent by the artist
Courtesy Max Protetch Gallery, New York

18 *Buddy* 1988
mixed media 23 1/2 x 9 1/2 x 7 1/2
Lent by the artist
Courtesy Max Protetch Gallery, New York

Steven CORTRIGHT

Born 1942, Long Beach, California
B.A., 1964, M.A., 1966, Stanford University, Palo Alto
Lives in Santa Barbara

19 *Untitled* 1988-1990
mixed media
tv: 40 x 25 x 24; chair: 38 x 23 x 30
Lent by the artist

Jack DOLLHAUSEN

Born 1941, Niagara Falls
B.A., Wisconsin State University, Whitewater, 1966
M.F.A., Washington State University, Pullman, 1968
Lives in Pullman

20 *Midwest Summer Night's Dream* 1977
mixed media 12 x 12 x 8
Lent by the artist

21 *Too Close* 1988
electronic circuitry 50 x 30
Lent by the artist

22 *Chance of Laws II* 1989
electronic circuitry 36 3/4 x 25 3/4 x 3 1/2
Lent by the artist

Marcel DUCHAMP

Born 1887, Blainville, Normandy;
died 1968, Neuille-sur-Seine, France
Académie Julien, 1904-1905

23 *Rotoreliefs* 1935-1953
mixed media 7 7/8 diameter
Collection David W. Bermant

Richard DUNLAP

Born 1939, Seattle
B.A., University of Washington, 1966
M.F.A., University of Washington, 1968
Lives in Santa Barbara

24 *History of Animals* 1989-1990
sound performance
Courtesy of the artist

R. M. FISCHER

Born 1947, New York City
B.A., Long Island University, Brookville, 1971
M.F.A., San Francisco Art Institute, 1973
Lives in New York City

25 *Snap, Crackle, Pop* 1986
brass, plastic, stainless steel, steel, electric lights 88 x 123 x 36
Collection Eli Broad Family Foundation, Santa Monica

26 *Amadeus* 1986
brass, aluminum, steel, plastic, electric lights 34 1/2 x 45 x 45 1/2
Collection Eli Broad Family Foundation, Santa Monica

Dan FLAVIN

Born 1933, New York City
Cathedral College of the Immaculate Conception, 1947-1952
Columbia University, New York, 1957-1959
Lives in Garrison, New York

27 *Monument for V. Tatlin* 1969
cool white fluorescent light, ed. 1/5 96 x 30 1/2
Collection The Museum of Contemporary Art, Los Angeles
Gift of Lenore S. and Bernard Greenberg

28 *Untitled (Fondly, to Helen)* 1976
blue, green and yellow fluorescent light 96 x 32
Collection Robert and Mary Looker

Bill FONTANA

Born 1947, Cleveland
B.A., The New School for Social Research, New York, 1970
Lives in Emeryville, California

29 *Elevated Soundings* 1990
site-responsive sound sculpture
Courtery of the artist and the David Bermant Foundation: Color, Light, Motion through a grant to the New Arts Foundation

Mimi GARRARD and James SEAWRIGHT

Garrard: Born Gastonia, North Carolina
B.A., Sweetbriar College, Sweetbriar, Virginia
Seawright: Born 1936, Jackson, Mississippi
B.A., University of Mississippi, 1957
Live in New York City

30 *Walking on Gravel* 1990
dance performance/light installation
Courtesy of the artists

Michael HARDESTY

Born 1952, Louisville, Kentucky
B.A., Eastern Kentucky University, Richmond, 1976
M.F.A., University of Arizona, Tucson, 1982
Lives in New York City

31 *As If* 1990
L.E.D. sign, incandescent blacklight, glow paint on acrylic
48 diameter
Lent by the artist
Courtesy Germans van Eck, New York

Michael HAYDEN

Born 1943, Vancouver
Lives in Los Angeles

32 *Coruscation* 1990
unidirectional diffraction grating; twelve units, each 66 x 36
Courtesy of the artist

Jenny HOLZER

Born 1950, Gallipolis, Ohio
B.F.A., Ohio University, Athens, 1972
M.F.A., Rhode Island School of Design, Providence, 1977
Lives in Hoosick Falls, New York

33 *More Survival* 1985
red diodes, 2" characters 5 1/2 x 60
Collection Martin and Sharleen Cohen

Rebecca HORN

Born 1944, Michelstadt, Germany
Hochschule für Bildende Kunste, Hamburg, 1964-1970
Lives in Bad König-Zell, West Germany

34 *Butterfly Machine* 1986
silver with parrot wings 49 1/2 x 6 x 9
Collection Martin Sklar

Robert IRWIN

Born 1928, Long Beach, California
Otis Art Institute, Los Angeles, 1948-1950
Jepson Art Institute, Los Angeles, 1951
Chouinard Art Institute, Los Angeles, 1951-1953
Lives in San Diego

35 *Untitled* 1969
acrylic cast disc, acrylic paint 54 diameter x 5
Collection Newport Harbor Art Museum, Newport Beach
Gift of Mr. and Mrs. John F. Kelsey

Joe JONES

Born 1934, New York City
Lives in Asolo, Italy

36 *Music Machine* 1976
plastic, wood, battery operated motor 14 x 13 x 12
Collection David W. Bermant

Howard JONES

Born 1922, Ilion, New York
B.A., Syracuse University
Lives in Arcadia, Missouri

37 *Sonic Six* 1967-68
brushed aluminum, sound, electronics 24 x 96 x 10
Lent by Brandyn Jones

Kristin JONES and Andrew GINZEL

Jones: Born 1956, Washington, DC
B.A., Rhode Island School of Design, Providence, 1979
M.F.A., Yale University, 1983
Ginzel: Born 1954, Chicago
Self-taught
Live in New York City

38 *Pacific Lagoon Project* 1990
fiberglass, polyethylene, sand, brass, stainless steel
30' x 50' diameter
Collection David W. Bermant

Jon KESSLER

Born 1957, Yonkers
B.F.A., State University of New York, Purchase, 1980
Whitney Museum Independent Study Program, 1981
Lives in Paris

39 *16 Rue Chanoinesse* 1989
mixed media 70 7/8 x 70 7/8 x 15 3/4
Courtesy Luhring Augustine, New York

40 *Blackout* 1989
glass, steel, aluminum, steel mesh, motor, lights 95 x 30 x 46
Courtesy Luhring Augustine Hetzler, Santa Monica

Milton KOMISAR

Born 1935, Nashville
B.A., Vanderbilt University, Nashville, 1958
M.A., University of California, Berkeley, 1964
Lives in Oakland

41 *The Philosopher's Geometry* 1988
computer controlled light sculpture 72 x 84 x 60
Collection David and Carol Becker

Harold LEHR

Born 1945, Boston
B.A., Art History, University of Rochester, 1967
M.F.A., Rhode Island School of Design, Providence, 1971
Lives in Larchmont, New York

42 *Water Sculptures* 1972
aluminum spheres, fiberglass poles, electric pump
264 x 168 x 192
Collection David W. Bermant

Julio LE PARC

Born 1928, Mendoza, Argentina
School of Fine Arts, Buenos Aires, 1942-1954
Lives in Paris

43 *Ceiling Light Sculpture* 1975
painted wood, stainless steel 48 x 2 1/4
Collection David W. Bermant

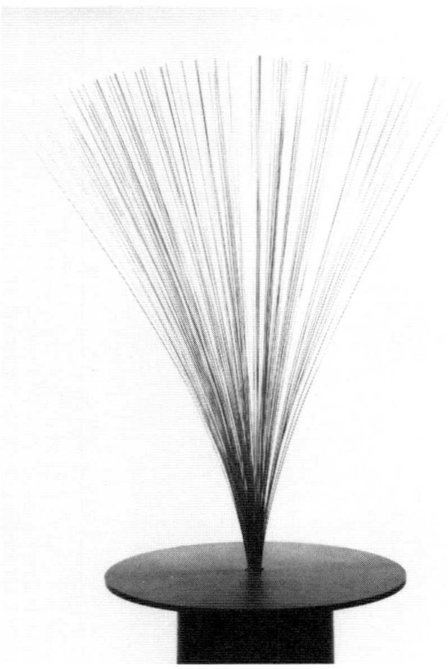

Len Lye
Fountain
1959

Evan LEWIS

Born 1958, New York City
B.F.A., The School of the Art Institute, Chicago, 1989
Lives in Chicago

47 *Quiet Storm* 1990
steel, aluminum, two units
20' diameter x 2 1/2' and 15' diameter x 2'
Lent by the artist

Len LYE

Born 1901, Christchurch, New Zealand;
died 1980, Warwick, New York

45 *Fountain* 1959
stainless steel rods, motor 85 x 72
Collection The Whitney Museum of American Art, New York
Gift of the Ford Foundation Purchase Program

Clyde LYNDS

Born 1936, Jersey City
Art Students League, New York, 1957-1958
Lives in Wallington, New Jersey

46 *Spectral Wind* 1974
fiber optics, formica case, motor 83 1/2 x 21 1/2 x 15 1/4
Collection David W. Bermant

47 *Radium Day* 1985
fiber optics, Plexiglas 36 x 26 x 7
Collection Art for Public Space, Inc., Harrison, New York

48 *Stele LXXV, Gnomon* 1989
fiber optics and concrete 94 x 15 x 14
Collection Art for Public Space, Inc., Harrison, New York

Heinz MACK

Born 1931, Lollar, Germany
Staatliche Kunstakadamie, Düsseldorf, 1950-1953
M.A., University of Cologne, 1956
Lives in Moenchengladbach, West Germany

49 *Light Dynamo No. 1* 1966
aluminum, electric motor 21 x 21 x 7 1/2
Collection David W. Bermant

David MAHLER

Born 1944, Plainfield, New Jersey
B.A., Concordia College, River Forest, Illinois, 1967
M.F.A., California Institute of the Arts, San Francisco, 1972
Lives in Seattle

50 *Listener in Residence & The Art of Tape* 1990
sound performance
Courtesy of the artist

Ron MALLORY

Born 1935, Philadelphia
B.A., Architecture, University of Florida, Gainesville, 1956
Lives in New York City

51 *Mercury Piece* 1965
motorized construction with mercury 12 3/4 x 12 3/4
Collection David W. Bermant

László MOHOLY-NAGY

Born 1895, Borsod, Hungary; died 1947, Chicago
Studied University of Budapest, 1913-1915

52 *Light Display—Black and White and Gray* 1930
film (video)
Courtesy The Museum of Modern Art, New York

Bruce NAUMAN

Born 1941, Fort Wayne
B.S., University of Wisconsin, Madison, 1964
M.F.A., University of California, Davis, 1966
Lives in Pecos, New Mexico

53 *Malice* 1980
neon and glass tubing 7 x 29 x 3
Private Collection

Rikuro OKAMOTO

Born 1943, Tokyo
B.E., Tokyo Institute of Technology, 1969
Lives in New York City

54 *Moving Rock No. 22* 1988
plastic resin, sand, aluminum, stainless steel, motor 33 x 47 x 81
Collection Art for Public Space, Inc., Harrison, New York

Dennis OPPENHEIM

Born 1938, Electric City, Washington
B.F.A., California College of Arts and Crafts, Oakland, 1964
M.F.A., Stanford University, Palo Alto, 1965
Lives in New York City

55 *Ghost Tantrums, Static Tremors* 1988-1989
flexi-mirror shields, silkscreen, steel, casters, electric train, track, wooden letters 144 x 72 x 36
Lent by the artist

56 *Four Spinning Dancers* (from *The Power Tools Series*) 1989
aluminum shelf, reflector wall, four buffing disk dolls, four electric drills 48 x 72 x 24
Lent by the artist

Eric ORR

Born 1939, Covington, Kentucky
Studied University of California, Berkeley; New School for Social Research; Ecole de Pataphysiques, Paris; University of Cincinnati
Lives in Los Angeles

57 *Prime Matter VIII* 1987
bronze, fire, water 114 x 9 x 8
Collection David W. Bermant

Nam June PAIK

Born 1932, Seoul, Korea
University of Tokyo, 1952-1956
Lives in New York City

58 *Participation TV* 1969
Philco color lite TV, microphones, amplifiers 30 x 31 3/4 x 24
Collection David W. Bermant

Liz PHILLIPS

Born 1951, Jersey City
B.A., Bennington College, Vermont, 1973
Lives in Astoria, New York

59 *Soundtable I and II (Wet and Dry Landscape)* 1989-1990
mixed media 44 x 44 x 32 1/2 and 44 x 44 x 12
Lent by the artist

Alan RATH

Born 1959, Cincinnati
B.S.E.E., Massachusetts Institute of Technology, Cambridge, 1982
Lives in Oakland

60 *Introspective Sculpture* 1988
aluminum, electronics, cathode ray tube 14 x 17 x 25
Collection David W. Bermant

61 *Desktop Breather* 1989
aluminum, electronics, speaker 16 x 12 x 8
Collection David W. Bermant

62 *Voyeur II* 1989
aluminum, steel, electronics, digital computer 83 x 25 x 68
Lent by the artist
Courtesy Dorothy Goldeen Gallery, Santa Monica

George RHOADS

Born 1926, Chicago
B.A., University of Chicago, 1945
Lives in Ithaca

63 *Goodtime Clock IV* 1984
mixed media, electric motor 72 x 72 x 36
Collection National Shopping Centers, Inc., Harrison, New York

64 *Wall Piece XVI* 1987
mixed media, electric motor 40 x 60 x 13 1/2
Collection Art for Public Space, Inc., Harrison, New York

65 *Lunaticks* 1989
wood, paper, batteries, clocks, motors 36 x 120
Collection Art for Public Space, Inc., Harrison, New York

76 *BeeTree III* 1990
stainless steel 96 x 66
Collection Art for Public Space, Inc., Harrison, New York

George RICKEY

Born 1907, South Bend, Indiana
B.A., Balliol College, Oxford, 1929
M.A., Balliol College, Oxford, 1941
Lives in East Chatham, New York and Santa Barbara

67 *Tidal V* 1963
stainless steel 14 x 74 x 14
Collection Barry and Gail Berkus

68 *Four Trapezoids as Two Rectangles II* 1985-1987
stainless steel 112 x 31 1/2
Collection Carl and Judy Schlosberg

Raúl RUIZ

Born 1941, Puerto Montt, Chile
Lives in Paris

70 *Expulsion of the Moors* 1989
Multi-media installation
Courtesy Institute of Contemporary Art, Boston

James SEAWRIGHT

Born 1936, Jackson, Mississippi
B.A., University of Mississippi, 1957
Lives in New York City

70 *House Plants* 1984
metal, plastic, digital computer 22 x 4 x 5
Collection Art for Public Space, Inc., Harrison, New York

71 *Mirror XVII* 1987
cast stone and mirrors 72 x 72 x 4
Collection Art for Public Space, Inc., Harrison, New York

Alejandro and Moira SIÑA

Alejandro: Born 1945, Santiago, Chile
M.B.A., University of Chile, 1973
M.I.T. Center for Advanced Visual Studies, 1973-1979
Moira: Born 1954, Portsmouth, Virginia
Live in Brookline, Massachusetts

72 *Corona Discharge* 1972-1976
water, light bulb, steel, electronics 25 x 15
Collection David W. Bermant

73 *Night Butterfly* 1980
neon, flexible steel rod, motor 120 x 72 x 72
Collection David W. Bermant

74 *Spinning Shaft* 1984
neon tubes, electric motor 67 x 128 x 37
Collection David W. Bermant

75 *Helicoil* 1987
neon, electronics, motor, cable 168 x 24 radius
Collection David W. Bermant

76 *Touch Plane* 1988
neon, electronics, wood 24 x 48 x 6
Lent by the artists

Jesus-Raphael SOTO

Born 1923, Ciudad Bolivar, Venezuela
Academia de Bellas Artes, Caracas, 1942-1947
Lives in Paris

77 *Rond Blanc* 1970
painted wood, wire 37 1/2 diameter
Collection David W. Bermant

Art SPELLINGS

Born 1950, Memphis, Tennessee
B.F.A., Indiana University, Indianapolis, 1974
Lives in Brooklyn

78 *Free Spirit* 1986
mixed media with motor 65 x 61 x 16
Collection National Shopping Centers, Inc., Harrison, New York

SURVIVAL RESEARCH LABORATORIES

Collective formed 1979, San Francisco
Mark Pauline, Artistic Director

79 *San Francisco Artspace* 1989
video of installation, directed by Jonathan Reiss and Leslie Asako Gladsjo
Lent by Survival Research Laboratories, San Francisco

Vasilli TAKIS

Born 1925, Athens, Greece
Fellow at Center for Advanced Visual Studies, 1968-1969
Lives in London

80 *Musicale* 1965-1968
metal, magnets, needles, electricity 49 1/4 x 14 3/4
Collection David W. Bermant

81 *Signal* (series 2, ed. no. 113) 1968
electronics, magnets 83 1/2
Collection David W. Bermant

Jean TINGUELY

Born 1925, Fribourg, Switzerland
Kunstgewerbeschule, Basel, 1941-1945
Lives in Essones, France

82 *Tokyo Gal* 1967
radio parts, flywheel, feather, motor 29 x 10 1/2 x 5 1/2
Collection David W. Bermant

Ted VICTORIA

Born 1944, Riverhead, New York
B.A., State University of New York, New Paltz, 1968
M.F.A., Rutgers University, 1970
Lives in New York City

83 *Hey Pete, Heads or Tails?* 1989
mixed media 24 x 36 x 5
Lent by the artist

84 *Fill It to the Brim* 1990
mixed media 10 x 7 x 20'
Collection David W. Bermant

85 *Bottoms Up* 1990
mixed media 10 x 7 x 20'
Collection David W. Bermant

Peter VOGEL

Born 1937, Freiburg, Germany
University of Freiburg
Lives in Freiburg

86 *Two Different Series* 1979
cybernetic relief 22 x 38
Collection David W. Bermant

Exhibitions Frequently Cited

1955

Le Mouvement, Galerie Denise René, Paris

1961

International Exhibition of Art in Motion, Stedelijk Museum, Amsterdam

1965

Art and Movement, Royal Scottish Academy, Edinburgh; Glasgow Art Gallery

Art and Movement, Musée de Tel Aviv

Group Zero, Gallery of Modern Art, Washington, DC

Kinetic and Object, Staatsgalerie, Stuttgart; Kunstverein, Karlsruhe

Kinetic Art from Krefeld, Gemeente Museum, The Hague; Stedelijk-van-Abbe Museum, Eindhoven

NUL, Stedelijk Museum, Amsterdam

Kinetic and Optic Art Today, Albright-Knox Art Gallery, Buffalo

Light and Movement, Kunsthalle, Bern; Palais des Beaux Arts, Brussels; Staatliche Kunsthalle, Baden Baden

Mouvement II, Galerie Denise René, Paris

Nove Tendencije III, Zagreb

1966

Directions in Kinetic Sculpture, University Art Museum, Berkeley and Santa Barbara Museum of Art

Light as a Creative Medium, Carpenter Art Center, Harvard University, Cambridge

Kunst, Licht, Kunst, Stedelijk-van-Abbe Museum, Eindhaven

9 Evenings, The Sixty-ninth Regimental Armory, New York

1967

Light/Motion/Space, Walker Art Center, Minneapolis, and Milwaukee Art Center

Light and Motion, Worcester Art Museum, Massachusetts

1968

Cybernetic Serendipity, Institute of Contemporary Art, London

The Machine as Seen at the End of the Mechanical Age, The Museum of Modern Art, New York; University of St. Thomas, Houston; San Francisco Museum of Art

The Magic Theater, Nelson Gallery of Art and Atkins Museum, Kansas City, Missouri

1969

Electric Art, UCLA Art Galleries, Phoenix Art Museum

1971

Art and Technology, Los Angeles County Museum of Art

1973

Sound/Sculpture, Vancouver Art Gallery, British Columbia

1978

Art of the Space Era, Huntsville Museum of Art, Alabama

1979

Sound, Los Angeles Institute of Contemporary Art; Project Studios 1, New York

1980

Für Augen und Ohren, Objekte, Installationen, Performances, Akademie der Kunste, Berlin, 1980

1981

Soundings, Neuberger Museum, State University of New York, Purchase

1982

Bruce Nauman: Neons, The Baltimore Museum of Art

Cork Marcheschi, The Minneapolis Institute of Arts

1984

The David Bermant Collection: Color, Light, Motion, Wadsworth Atheneum, Hartford

1985

Lumières, Je pense à vous, Centre de Creation Industrielle, Paris

Modern Machines: Recent Kinetic Art, Whitney Museum of American Art at Philip Morris, New York

1986

Engaging Objects: The Participatory Art of Mirrors, Mechanisms and Shelters, The Clocktower, Institute for Art and Urban Resources, Long Island City, New York

Lumières, Perception—Projection, Les Cent Jours d'Art Contemporain, Centre International d'Art Contemporain, Montreal

1986-1987
Jenny Holzer: Signs, Des Moines Art Center, Iowa

1987
Art Ware, Kunst und Electronik, Hannover- Messe CeBIT; Landesmuseum Volk und Wirtschaft, Düsseldorf

Digital Visions, Everson Museum of Art, Syracuse; I.B.M. Gallery, New York

Jack Dollhausen, Redmann Galerie + Kunsthandel, Berlin

Eccentric Machines, John Michael Kohler Arts Center, Sheboygan, Wisconsin

Klangsculpteren/Sound Sculptures by Bill Fontana, Museum Ludwig, Köln; San Francisco Museum of Modern Art

1988
Interaction, Light, Sound, Motion, Aldrich Museum of Contemporary Art, Ridgefield, Connecticut

Ted Victoria, Opere di Luca, Museo Fortuny, Venice

1989
The Drowned World: Waterworks, The Institute for Contemporary Art, P.S.I. Museum, Long Island City, New York

Jenny Holzer, Solomon R. Guggenheim Museum, New York

Signals, QCC Art Gallery, Queensborough Community College, City University of New York, Bayside, New York

The Techno/Logical Imagination: Machines in the Garden of Art, The Minneapolis College of Art and Design

Wonderland of Science Art—Invitation to Interactive Art, Kanagawa Scientific Engineering Academy, Japan

1990
Michael Hardesty, Pittsburgh Center for the Arts

Selected Reading List

Exhibition Catalogues

Andrews, Oliver. *Electric Art*. Los Angeles: UCLA Art Galleries, 1969.

Armstrong, Richard. *The Sculpture of R. M. Fischer*. (brochure) New York: Whitney Museum of American Art, 1984.

Atkinson, Tracy. *The David Bermant Collection: Color, Light, Motion*. Hartford: Wadsworth Atheneum, 1984.

Bard, Perry. *Signals*. Bayside, New York: QCC Art Gallery, Queensborough Community College of the City University of New York, 1989.

Cameron, Dan. *Michael Hardesty*. Pittsburgh: Pittsburgh Center for the Arts, 1990.

Coe, Ralph T. *The Magic Theater, Art Technology Spectacular*. Kansas City, Missouri: Nelson Gallery of Art, 1970

Cubbs, Joanne. *Eccentric Machines*. Sheboygan, Wisconsin: John Michael Kohler Arts Center, 1987.

Delahanty, Suzanne and Dore Ashton. *Soundings*. Purchase, New York: Neuberger Museum, 1981.

Dollhausen, Jack. *Jack Dollhausen*. Berlin: Galerie Redmann, 1987.

Finkelpearl, Tom. *Engaging Objects: The Participatory Art of Mirrors, Mechanisms and Shelters*. Long Island City, New York: The Clocktower, Institute for Art and Urban Resources, 1986.

_____. *P.U.L.S.E.* New York: The Clocktower, Institute for Art and Urban Resources, 1987.

_____. *The Drowned World: Waterworks*. Long Island City, New York: The Institute for Contemporary Art, P.S. 1 Museum, 1989.

Flavin, Dan, Sabine Aschendorf, Madeleine Deschamps and Jochen Potter. *Dan Flavin/New Uses for Fluorescent Light*. Baden Baden: Staatliche Kunsthalle, 1989.

Fontana, Bill. *Klangskulpteren/Sound sculptures*. Köln, Musuem Ludwig, 1987.

Galloway, David. *Artware, Kunst und Electronik*. Hanover: Hanover-Messe, 1987.

Gascon, France. *Roland Brener, Michel Goulet*. Montreal: Musee d'art contemporain de Montreal, 1988.

Goodman, Cynthia. *Digital Visions*. New York: Harry N. Abrams and Everson Museum of Art, 1987.

Gosselin, Claude; Danielle Roy, Luc Courchesne and Robert White. *Lumières, Perception - Projection*. Montreal: Centre International d'Art Contemporain, 1986.

Grunfeld, Jean François, ed. *Lumiere, Je pense à vous*. Paris: Centre Georges Pompidou and Editions Herme, 1985.
Hanhardt, John G. *Nam June Paik*. New York: Harry N. Abrams and Everson Museum of Art, 1982.

Hopps, Walter. *Marcel Duchamp: A Retrospective Exhibition*. Pasadena: Pasadena Art Museum, 1963.

Hulten, K. G. Pontus. *The Machine as Seen at the End of the Mechanical Age*. New York: The Museum of Modern Art, 1968.

Hunter, Sam; Gyorgy Kepes, Jonathan Benthall, Vilem Flusser, Tsai Wen-Ying and John Woodford. *Cybernetic Art of Tsai Wen-Ying*. Taipei: National Museum of History and Taichung: Taiwan Museum of Art, 1989.

Krickeberg, Dieter, Siegfried Wendel, Wolf D. Kuhhnelt, Juan Allende-Blin, Billy Kluver and René Block. *Für Augen und Ohren*. Berlin: Akademie der Kunste, 1980.

Lipman, Jean. *Calder's Universe*. Philadelphia: Running Press in cooperation with Whitney Museum of American Art, 1989.

Lubowsky, Susan. *Modern Machines: Recent Kinetic Art*. New York: Whitney Museum of American Art at Philip Morris, 1985.

McClintic, Miranda. *Clyde Lynds: Stelae*. Washington, D.C.: Wallace Wentworth Gallery, 1987.

Mescola, Sandro and Ann-Sargent Wooster. *Opere di Luce*. Venice: Museo Fortuny, 1988.

O'Donnell, Ellen M. and Miranda McClintic. *Interaction: Light, Sound, Motion*. Ridgefield, Connecticut: Aldrich Museum of Contemporary Art, 1988.

Osterwald, Tilman and Andreas Vowinkel. *Alice Aycock, Retrospective of Projects and Ideas, 1972-1983*. Stuttgart: Wurttembergischer Kunstverein Stuttgart, 1983.

Richardson, Brenda. *Bruce Nauman: Neons*. Baltimore: The Baltimore Museum of Art, 1983.

Selz, Peter, and George Rickey. *Directions in Kinetic Sculpture*. Berkeley: University Art Museum, 1966.

Sharp, Willoughby. *Light/Motion/Space*. Minneapolis and Milwaukee: Walker Art Center and Milwaukee Art Center, 1967.

Shulman, Leon. *Light and Motion*. Worcester, Massachusetts: Worcester Art Museum, 1967.

Sieling, Neil. *The Techno/Logical Imagination: Machines in the Garden of Art*. Minneapolis: Intermedia Arts, Minnesota, and Minneapolis College of Art and Design, 1989.

Simon, Joan. *Jenny Holzer: Signs*. Des Moines, Iowa: Des Moines Art Center, 1986.

Smith, Robert and Bob Wilhite. *Sound*. Los Angeles: Los Angeles Institute of Contemporary Art, 1979.

Tuchman, Maurice. *Report on the Art and Technology Program of the Los Angeles County Museum of Art*. Los Angeles: Los Angeles County Museum of Art, 1971.

Waldman, Diane. *Jenny Holzer*. New York: Solomon R. Guggenheim Museum, 1989.

Warren, Lynne. *Jon Kessler*. Chicago: Museum of Contemporary Art, 1986.

Wood, Carolyn. *Art of the Space Era*. Huntsville, Alabama: Huntsville Museum of Art, 1978.

Books

Burnham, Jack. *Beyond Modern Sculpture*. New York: George Braziller, Inc., 1968.

Hanhardt, John, ed. *Video Culture*. Rochester, New York: Gibbs M. Smith, Inc., Peregrine Smith Books in association with Visual Studies Workshop, 1986.

Honisch, Dieter. *Mack Sculptures, 1953-1986*. Düsseldorf: Econ, 1987.

Jardine, Alice. "Of Bodies and Technologies." Hal Foster, ed. *Discussions in Contemporary Culture* 1, Seattle: Bay Press and Dia Art Foundation, 1987.

Kepes, Gyorgy, ed. *The Nature and Art of Motion*. Vision + Value series. New York: George Braziller, 1965.

Krauss, Rosalind E. *Passages in Modern Sculpture*. New York: The Viking Press, 1977.

Kuenzli, Rudolph E. and Francis N. Naumann, eds. *Marcel Duchamp, Artist of the Century*. Cambridge: The MIT Press, 1989.

Moholy-Nagy, László. *Vision in Motion*. Chicago: P. Theobald, 1947.

Nelson, June Kompass. *Harry Bertoia, Sculptor*. Detroit: Wayne State University Press, 1970.

Reichardt, Jasia, ed. *Cybernetic Serendipity, The Computer and the Arts*. London: The Studio International, 1968.

Rickey, George. *Constructivism, Origins and Evolution*. New York: George Braziller, 1967.

Wodiczko, Krzysztof. "Strategies of Public Address: Which Media? Which Public?" Hal Foster, ed. *Discussions in Contemporary Culture* 1, Seattle: Bay Press and Dia Art Foundation, 1987.

Articles

Baschet, François and Bernard Baschet. "Sound Sculpture," *Leonardo,* 1987, 107-14.

Bell, Bill, E. Theodore Parks and Robert B. Post. "Elusive Imagery of the Lightstick," *Leonardo,*1986, 3-10.

Boettger, Suzann. "Milton Komisar, New Langton Arts," *Artforum,* November 1985, 111-2.

Cameron, Dan. "Horn's Dilemma: The Art of Rebecca Horn," *Arts Magazine*, November 1987, 72-5.

Celant, Germano. "Rebecca Horn, Dancing on the Egg," *Artforum,* October 1984, 48-55.

Cooke, Lynne. "Anamnesis," *Artscribe International*, May 1988, 50-4.

Ferguson, Bruce. "Wordsmith, An Interview with Jenny Holzer," *Art in America*, December 1986, 108-27.

Fischer, R. M. "What I Believe," *Cimaise,* May/July 1984, 37-48.

Fontana, Bill. "The Relocation of Ambient Sound," *Leonardo,* 1987, 143-7.

Foster, Hal. "Subversive Signs," *Art in America*, November, 1982, 88-92.

Harrison, Charles. "The Sculpture of Roland Brener," *Artforum,* June 1969, 32-5.

Horn, Rebecca. "Collaboration," *Parkett* 13, 1987.

Jouffroy, Alain. "le mouvement des inventions et des machines," *XXe Siecle*, December 1976, 96-103.

Kossy, Barbara. "Listening to Berkeley," *Artweek,* 23 November 1985, 3.

Kurtz, Bruce. "Paikvision," *Artforum,* October 1982, 52-55.

Kuspit, Donald. "Aycock's Dream Houses," *Art in America*, September 1980, 84-7.

Le Parc, Julio. "The Laws of Chance," *Studio International*, January 1973, 2-3.

McEvilley, Thomas. "Negative Presences in Secret Spaces, The Art of Eric Orr," *Artforum,* Summer 1982, 58-66.

Piene, Nan R. "Light Art," *Art in America*, May/June 1967, 24- 47.

Popper, Frank. "The Luminous Trend in Kinetic Art," *Studio International*, February 1967, 173-6.

Reichardt, Jasia. "E.A.T. and after," *Studio International*, May 1968, 236-7.

Rickey, George. "The New Tendency (Nouvelle Tendance-Recherche Continuelle)," *Art Journal*, Summer 1964, 272-9.

_____. "Kinesis Continued," *Art in America*, December 1965, 45- 55.

Rose, Barbara. "Kinetic Solutions to Pictorial Problems - The Films of Man Ray and Moholy-Nagy," *Artforum,* September 1971, 68- 73.

Schwartz, Barry N. "Art and Engineering," *Arts Magazine*, Summer 1968, 17.

Selz, Peter. "Milton Komisar's `Nisus'," *Arts Magazine*, June 1981, 85-7.

Siegel, Jeanne. "Jenny Holzer's Language Games," *Arts Magazine*, December 1985, 64-8.

Simon, Joan. "Breaking the Silence," *Art in America*, September 1988, 141-8, 203.

Staber, Margit. "Heinz Mack - Einige Festellungen Zu Seinem Werk," *Art International*, Summer 1969, 17-20, 76.

Thorburn, Ray. "Ray Thorburn Interviews Len Lye," *Art International*, April 1975, 64-8.

Tsai, Eugenie. "A Tale of (at least) Two Cities," *Arts Magazine*, June 1982, 134-41.

Wood, Steve. "An Interview with Dennis Oppenheim," *Arts Magazine*, June 1981, 133-37.

Wooster, Ann-Sargent. "Art Sounds," *Art in America*, February 1982, 116-25

_____. "Why Don't They Tell Stories Like They Used To?" *Art Journal*, Fall 1985, 204-12.

University Art Museum